ARTIFICIAL INTELLIGENCE IN AGRICULTURE

About the Authors

Dr. Rajesh Singh is currently associated with Lovely Professional University as Professor with more than Sixteen years of experience in academics. He has been awarded as gold medalist in M.Tech from RGPV, Bhopal (M.P) India and honors in his B.E from Dr. B.R. Ambedkar University, Agra (U.P), India. His area of expertise includes embedded systems, robotics, wireless sensor networks and Internet of Things. He has been honored as keynote speakers and session chair to international/national conferences, faculty development programs and workshops. He has **hundred and fifty two patents** in his account. He has published more than **hundred research papers** in referred journals/ conferences and **twenty four books** in the area of Embedded Systems and Internet of Things with reputed publishers like CRC/Taylor & Francis, Narosa, GBS, IRP, NIPA, River Publishers, Bentham Science and RI publication. He is editor to a special issue published by AISC book series, Springer in 2017 & 2018 and IGI global in 2019. Under his mentorship students have participated in national/international competitions including "**Innovative Design Challenge competition**" by Texas and DST and **Laureate award of excellence in robotics engineering**, Madrid, Spain in 2014 & 2015. His team has been the winner of "**Smart India Hackathon-2019**" hardware version conducted by MHRD, Government of India for the problem statement of Mahindra & Mahindra. Under his mentorship students team got "**InSc award 2019**" under students projects program. He has been awarded with "**Gandhian Young Technological Innovation (GYTI) Award**", as **Mentor** to "**On Board Diagnostic Data Analysis System-OBDAS**", Appreciated under "Cutting Edge Innovation" during Festival of Innovation and Entrepreneurship at **Rashtrapati Bahawan**, India in 2018. He has been honored with "**Certificate of Excellence**" from 3rd faculty branding awards-15, Organized by EET CRS research wing for excellence in professional education and Industry, for the category "Award for Excellence in Research", 2015 and young investigator award at the International Conference on Science and Information in 2012.

Dr. Anita Gehlot is currently associated with Lovely Professional University as Associate Professor with more than twelve years of experience in academics. Her area of expertise includes embedded systems, wireless sensor networks and Internet of Things. She has been honoured as keynote speakers and session chair to international/national conferences, faculty development programs and workshops. She has

hundred and thirty two patents in her account. She has published more than seventy research papers in referred journals/conferences and **twenty four** books in the area of Embedded Systems and Internet of Things with reputed publishers like CRC/Taylor & Francis, Narosa, GBS, IRP, NIPA, River Publishers, Bentham Science and RI publication. She is editor to a special issue published by AISC book series, Springer in 2018 and IGI global in 2019. She has been awarded with **"certificate of appreciation"** from University of Petroleum and Energy Studies for exemplary work. Under her mentorship students team got "InSc award 2019" under students projects program. She has been awarded with "Gandhian Young Technological Innovation (GYTI) Award", as Mentor to "On Board Diagnostic Data Analysis System-OBDAS", Appreciated under "Cutting Edge Innovation" during Festival of Innovation and Entrepreneurship at Rashtrapati Bahawan, India in 2018.

Mahesh Kumar Prajapat is currently pursuing his Master's Degree in Computer Science and Engineering from Lovely Professional University. His major research area is Artificial Intelligence by which he has 16 patents in His account. He has won AIU Anveshan 2020 a research competition with 3rd position. He is a founder of startup Orphorhard which works for the development of orphans and underprivileged children around the country. He has done specialization courses which include deep learning, data science from Coursera, and IBM. he is always Acquiring broaden Horizon, Much observant in nature, perceiving self-generated ideologies and implementing that same and Love to explore different fields with full zest.

Bhupendra Singh is Managing director of Schematics Microelectronics and provides Product design and R&D support to industries and Universities. He has completed BCA, PGDCA, M.Sc. (CS), M.Tech and has more than eleven years of experience in the field of Computer Networking and Embedded systems. He has has published twelve books in the area of Embedded Systems and Internet of Things with reputed publishers like CRC/Taylor & Francis, Narosa, GBS, IRP, NIPA, River Publisher, Cambridge Scholar, Bentham Science and RI Publication.

ARTIFICIAL INTELLIGENCE IN AGRICULTURE

Rajesh Singh
Lovely Professional University
Jalandhar, Punjab, India

Anita Gehlot
Lovely Professional University
Jalandhar, Punjab, India

Mahesh Kumar Prajapat
Lovely Professional University
Jalandhar, Punjab, India

Bhupendra Singh
Schematics Microelectronics, India

CRC Press
Taylor & Francis Group
Boca Raton London New York

CRC Press is an imprint of the
Taylor & Francis Group, an **informa** business

NEW INDIA PUBLISHING AGENCY
New Delhi-110 034

First published 2022
by CRC Press
2 Park Square, Milton Park, Abingdon, Oxon, OX14 4RN

and by CRC Press
6000 Broken Sound Parkway NW, Suite 300, Boca Raton, FL 33487-2742

© 2022 New India Publishing Agency

CRC Press is an imprint of Informa UK Limited

The right of Rajesh Singh et al. to be identified as authors of this work has been asserted in accordance with sections 77 and 78 of the Copyright, Designs and Patents Act 1988.

Print edition not for sale in South Asia (India, Sri Lanka, Nepal, Bangladesh, Pakistan or Bhutan).

British Library Cataloguing-in-Publication Data
A catalogue record for this book is available from the British Library

Library of Congress Cataloging-in-Publication Data
A catalog record has been requested

ISBN: 978-1-032-15810-5 (hbk)
ISBN: 978-1-003-24575-9 (ebk)

DOI: 10.1201/9781003245759

Preface

This book is a platform for anyone who wishes to explore Artificial Intelligence in the field of agriculture from scratch or broaden their understanding and its uses.

This book offers a practical, hands-on exploration of Artificial Intelligence, machine learning, deep Learning, computer vision and Expert system with proper examples to understand. This book also covers the basics of python with example so that anyone can easily understand and utilize artificial Intelligence in agriculture field.

This book is divided into two parts wherein first part talks about the artificial intelligence and its impact in the agriculture with all its branches and their basics. The second part of the book is purely implementation of algorithms and use of different libraries of machine learning, deep learning and computer vision to build useful and sightful projects in real time which can be very useful for you to have better understanding of artificial Intelligence.

After reading this book, you'll have an understanding of what Artificial Intelligence is, where it is applicable, and what are its different branches, which can be useful in different scenarios. You'll be familiar with the standard workflow for approaching and solving machine-learning problems, and you'll know how to address commonly encountered issues. You'll be able to use Artificial Intelligence to tackle real-world problems ranging from crop health prediction to field surveillance analytics, classification to recognition of species of plants etc.,

Authors are thankful to the publisher for the support and encouragement to write this book.

Authors

Contents

Contents

PART A: FUNDAMENTAL OF ARTIFICIAL INTELLIGENCE

Chapter1

Artificial Intelligence

The artificial intelligence (AI) world is quickly growing as it reaches several different industries. You currently will see AI used for many different purposes from agriculture to automotive; with the passage of time you will see even more of it. Agriculture is one of the most important branches of the IT industry. Agriculture is an important industry and is a big part of our economy's base. In terms of annual sales, the agricultural industry contributes to our economy almost 17%.AI is becoming a technical advancement that boost and protects crop yield, as climate change and populations rise. Here in this chapter basics of Artificial Intelligence have been discussed how it can be related to agriculture.

1. Introduction

Artificial Intelligence has become most prominent in current generation. Every field is utilizing Artificial Intelligence for their better results and which also result in saving a lot of time. As Artificial intelligence is boosting different sectors to increase the productivity and efficiency and its utilization across different sectors like in the field of CIVIL Artificial Intelligence can be used to analyze the structures of buildings, dams etc., and Artificial Intelligence solutions are assisting in so many ways to overcome the traditional challenges in every field.

I. What is Artificial Intelligence

Artificial Intelligence is one of the most prominent field of Computer Science Engineering which attempts to redefine the tasks which are carried out by humans in their daily life with better results and maximum accuracy. Artificial Intelligence is actually meant to ease the work of human where the word artificial states the work done without any human interference and intelligence states the capabilities like human brain to perform tasks. The field artificial Intelligence has given so many definitions from time to time which can be typically stated as

1. The Engineering and Science come together to make machine Intelligent.
2. To make a machine intelligent which can imitate the human behavior
3. The branch of Computer Science which can perform human like activities which can include thinking, learning etc.,
4. An area of Computer Science which makes machine to act and behave like humans.

Artificial intelligence enables and increase the machine capabilities and makes them to comprehend facts, events, analyze the situations and understand to derive the knowledge from surrounding and to act upon them to solve the problem which can lead to better results then humans. The primary focus of Artificial Intelligence is always been to develop a machine which can phycological and to simulate the behavior of humans through constructing the thoughts like human from any surrounding or situations.

II. Introduction of Artificial Intelligence in Agriculture

Agriculture has been one of the oldest professions in the world. Humanity from their early ages to current generation has come to a long way in development of agricultural practices and its techniques in production of food humanity consumes in their daily lives. As it has seen from past generations to till current generation there always been an increase in population which leads to increase in consumption of food and decrease in production of it by 20%.

As Agricultural Production is decreasing day by day and from generation to generations, Artificial Intelligence can help us to automate the production of food crops in agricultural fields without any human interference which will lead into increase of food production. Artificial Intelligence solutions can perform tasks like analyzing the field for crop harvesting, monitoring of crops and soil conditions from time to time, it can forecast the weather and predict it for future purpose etc., using artificial intelligence in agriculture can leverage the assistance to the farmers and improve their farming techniques by serving them with correct data as per requirement. Later on discuss about types of AI and its different branches which can have a huge impact on Agriculture.

III. Stages and types of Artificial Intelligence

Artificial intelligence has different stages which can also be said to types of artificial intelligence which are

1. Artificial Narrow Intelligence
2. Artificial General Intelligence
3. Artificial Super Intelligence

These three types can said to be how AI can evolves, and now lets see each of them separately

Artificial Narrow Intelligence (ANI): It can also be called as weak AI, ANI can perform only the narrow-defined tasks which don't have cognitive and thinking ability, these are just the predefined functions which are already feeded into the system.

For Example, Siri, Alex, Self-Driving cars, Alpha Go, Sophia the humanoid etc., are the ANI we have developed so far.

Artificial General Intelligence (AGI): it can also be called as strong AI, AGI is the second stage of evolution of Artificial Intelligence which posses the ability to think and make decisions just like human. Currently we haven't developed such system which can said to the perfect example for artificial General Intelligence system.

Strong AI is by far considered as threat to human existence by so many researcher's and scientist and even by technology giants like Google, Microsoft, Tesla etc., The complete development of an Artificial Intelligence System could spell the end to an humanity and by far human is not capable of competing such system which is most powerful in terms of thinking and intelligence.

Artificial Super Intelligence (ASI): It can be stage of Artificial Intelligence where the capacity of AI system will pass all the limitations and can suppress the human beings. ASI is current now only a hypothetical situation which can happen in near future if mankind continues their advancement in development of Artificial Intelligence.

As we have seen the stages of evolution of artificial intelligence let see some of the artificial intelligence system which are purely based on the functions they perform. They are

1. **Reactive Machine Artificial Intelligence:** This type of AI system takes the input as current data and perform the task based on current situations. Reactive AI machine system cannot analyze and predict the future task or their actions of any particular situation. These systems always perform from narrow range predefined tasks.

2. **Limited Memory AI:** These AI system mostly based on the data they have stored and usually perform the task by studying the previously performed task and also perform the prediction of future task by the machine.

3. **Theory of Mind AI:** These systems are more advanced type of AI which is used to speculate and play a major role in terms of psychology. These AI systems always performs the tasks related to Emotional Intelligence and the thoughts which can be comprehended in any situation of an individuals.

4. **Self-Aware AI:** These System so far most advanced and fall under last stage of artificial Intelligence where AI system consist of consciousness and become self-aware of everything. In the nearby future these can be achievable.

As we seen the different stages and types of Artificial Intelligence system, but apart from this these systems are made by using several technologies which are considered to be as a branch of artificial intelligence. They are

- Machine Learning
- Deep Learning
- Computer Vision
- Data Science
- Big Data

These branches of artificial Intelligence have been massively utilized in agriculture filed to improve production and bring efficiency in harvesting. We will discuss these things in later chapters briefly and their implementations with some projects.

IV. Applications of AI in Agriculture

Artificial Intelligence has many applications and can be used in so many ways in the field of agriculture. As Agricultural fields provides us so much data in terms of temperature, precipitation, solar radiation and wind speed which can be analyzed and help us to benefit in terms of harvesting from agricultural fields. The best part in implementing artificial intelligence in agriculture is that it will not remove the jobs of farmers but it will help the farmers to improve their process of harvesting in the fields.

Artificial intelligence can have best application in agriculture, some of them are like:

1. **Artificial Intelligence Agriculture Bots:** AI Enabled Agriculture bots can assist the farmers to find the more efficient ways to find better harvesting techniques in any type of agricultural field and guide the farmer to protect their crops from weeds. These bots can also help in the agriculture fields in harvesting the crops at larger volume and at faster pace than any human laborer's in this way farmers can harvest more crops in less time. By implementing computer vision with the bots, it can also detect the several diseases into the crops and can spay pesticides at correct time.

2. **Monitoring crop and soil health:** Artificial Intelligence can be utilized in an efficient way to monitor and identify the defects and nutrients deficiencies in the soil. With the help of computer vision's image recognition approach systems can be developed which can identify the defectness in crop and using the deep learning applications flora patterns can be analyzed in agriculture. These type of AI Enabled applications will make easy to understand the different parts of crops, their defects, plant pests and its disease.

3. **Agricultural Expert System:** Expert System can be best for farmers in providing proper guidance. Expert System are by default are the system which have expert level knowledge stored in them which is being gathered from different experts of that field and which is based on different situations and practices which are being performed by experts.

4. **Forecasted Weather data:** Artificial Intelligence in an advance way can help the farmers to make them updated regarding the sudden change in weather which will help them to protect their crop from hazardous situation. The Analysis done by the system can help the farmers to take precautions (need to think from here).

5. **AI with Drones:** Drone Technology helps the farmers to improve their crop yields and reduce the cost. User pre programs the drone's route and once it is deployed into the field which can have the aerial technology enabled with various features like capturing live image data of crops, spraying pesticides on disease detection at particular crops in the field etc., Drones can be programmed for any particular tasks too related to the agricultural field like analyzing the field from time to time and scanning it for any infected or diseased crops. Camera's in the drone which are enabled with imaging technology can help with overall field management by providing real time crop requirement of water, fertilizers, pesticides etc., Machine learning can be used in the proper analysis of crop and soil condition by providing us proper insight of crop and its requirements which can increase the production of crop on the field.

6. **Indoor Harvesting with AI:** Indoor harvesting has become a trend in current time where people can harvest different crops in small area by providing proper artificial lighting according to its requirements. By Using Artificial Intelligence, indoor harvesting will become more simpler and easier because it can analyze everything in a closed environment and can automate the process of harvesting either it can be hydroponics or aquaponics.

V. Advantages of AI in Agriculture

Artificial Intelligence has many advantages in different field and here it has been discussed about the advantages in agriculture field which are

- Artificial Intelligence (AI) has provided us more efficient methods to produce, harvest and can guide us to sell essential crops by analyzing the market values.
- AI can be implemented to check crop defectiveness and to improve harvest of healthy crops.
- The increase in technology utility based on artificial Intelligence can strengthen the agriculturally based business and markets to run more efficiently.
- AI can help to improve the crop management practices and its harvesting.
- AI based solutions can be used to solve the various traditional challenges which are being faced by farmers like climate change, an infestation of pests and weeds thar reduces yielding of food crops.

Summary

As so far discussed in this chapter regarding artificial Intelligence and its types narrowing down its advantages in agriculture. It is also discussed that artificial Intelligence can lead to increase in yield and protection of crops while practicing agriculture in the fields in real time which can be very useful to farmers nowadays. Due to advancement in technology farmers should also an advancement in agricultural techniques to increase their crop production and saving time in which artificial Intelligence can play a very big role.

Chapter 2

Learning Python for Artificial Intelligence

As we have seen how artificial Intelligence can be useful in agriculture but the nest question always come in mind is that from where and how we can start to learn and what are the model of learning an artificial Intelligence. As Artificial Intelligence is a huge part of Computer Science which means we need an language as medium to perform it and we always prefer Python for that which is very dynamic and easy to learn. In this chapter we will learn about python basics so that in further implementation of artificial Intelligence it will be helpful to us.

I. Introduction

Python is a remarkable language that can claim to be easy and strong. You will be amazed to see how simple it is not to focus on the language you are programming, but on solving the problem. Guido van Rossum was the creator of the Python language which is named after the BBC show "Monty Python's flying Circus".

Python is formally introduced as:

"Python is an easy to learn, powerful programming language. It has efficient high-level data structures and a simple but effective approach to object-oriented programming. Python's elegant syntax and dynamic typing, together with its interpreted nature, make it an ideal language for scripting and rapid application development in many areas on most platforms."

II. Features f Python

- **Simple:** The Python language is simple and minimalist. Reading a good Python program almost feels like reading English, but it's really tight! Python is one of its biggest assets, its pseudo-code existence. This helps you to focus instead of the language itself on the problematic solution.

- **Easy to Learn:** As you'll see, Python can be started very quickly. As already described, Python has an extremely simple syntax.

- **Free and open source:** A FLOSS (Free and Open Source Software) example is Python. You can share free copies, read the source code of the software, make significant changes, and use some of it in free new programs. FLOSS is based on an intelligence-sharing network idea. Which is one of the factors Why Python is developed and is continually changed.

- **High Level Language:** You never have to bother about small information such as the management of memory use in your program when you make programs in Python, etc.

- **Portable:** Python has been adapted to (i.e., updated to) several sites because of its existence as an open source. Either of these systems will operate in all your Python programs without any modifications if you are diligent enough to avoid system-dependent functions. Python can also be used for GNU/Linux, Windows, FreeBSD, Macintosh, Solaris, OS/2, Amiga, AROS, AS/400, BeOS, OS/390, z/OS, Palm OS, Windows CE and PocketPC.

  You can also make games for your machine, iPhone, iPad or Android with a platform like Kivy.

- **Extensible:** You could code some part of a program in c or c++ and then use it from a program called your Python if you need a vital part of code to run very quickly or you want any algorithm not to be opened.

- **Embeddable:** To offer users scripting capabilities for your software, you can incorporate Python in your C/C++ programs.

- **Interpreted:** A compiled program such as C or C++ is translated from the source language, i.e., C or C++ in a programming language (binary code i.e., 0s and 1s) with a compiler containing different flags and options The Linker/Loader software copies and begins running this program from hard disk to memory. On the other hand, Python does not require binary compilation. The software is run by the source code directly. Python dynamically transforms the source code into an intermediate type called bytecodes, then translates it into the computer's native language and then executes it. In fact, it's much easier to use Python because you don't have to worry if you compile the software to ensure the correct libraries are attached and loaded, etc. This makes your Python program more portable, as you can only copy your Python to a different machine and it just functions!

- **Object Oriented:** Python facilitates procedure-oriented programming as well as object-oriented programming. In procedure-oriented languages, the software is constructed around procedures or functions which are nothing but generic pieces of programs. The software is designed around objects in object-oriented languages that combine data and functionality. In comparative with major languages such as C++ or Java, Python has a good but simplified way to do OOP.

- **Extensive libraries:** The basic library of Python is immense. It will help you make various things with regular expressions, document creation, unit analysis, threading, data bases, web browsers, CGI, FTP, email, XML, XML-RPC, HTML, WAV, encryption, GUI and other system-dependent objects. All this should be considered anywhere Python is installed. This is known as the Theory of Python, including batteries.

  Besides the main library, the Python Package Index also provides many other high-quality libraries.

III. Installation

A large number of platforms provide Python distribution. The binary code for your application must only be downloaded and Python enabled.

You need a C compiler to compile the source code by hand if the binary code of your platform is not usable. In order to make the choice of features required during your installation, you can compile the source code more flexibly.

A short picture of Python installation across different platforms is given here –

- Unix and Linux Installation
  - Follow these steps to install Python on Unix/Linux machine.
  - Open a Web browser and go to https://www.python.org/downloads/.
  - Follow the link to download zipped source code available for Unix/Linux.
  - Download and extract files.
  - Editing the Modules/Setup file if you want to customize some options.
  - run./configure script
  - make
  - make install
  - This installs Python at the standard location /usr/local/bin and its libraries at /usr/local/lib/pythonXX where XX is the version of Python.
- **Windows Installation**
  - Visit https://www.python.org/downloads/ and download the latest version. At the time of this writing, it was Python 3.7 or greater version The installation is just like any other Windows-based software.
  - Note that if your Windows version is pre-Vista, you should download Python 3.4 only as later versions require newer versions of Windows.
  - CAUTION: Make sure you check option Add Python 3.7 or greater version to PATH.
  - To change install location, click on Customize installation, then Next and enter C:\python35 (or another appropriate location) as the install location.
  - If you didn't check the Add Python 3.7 or greater version PATH option earlier, check Add Python to environment variables. This does the same thing as Add Python 3.7 or greater version to PATH on the first install screen.
  - You can choose to install Launcher for all users or not, it does not matter much. Launcher is used to switch between different versions of Python installed.
  - If your path was not set correctly (by checking the Add Python 3.7 or greater version Path or Add Python to environment variables options), then follow the steps in the next section (DOS Prompt) to fix it. Otherwise, go to the Running Python prompt on Windows section in this document.

  Note: For people who already know programming, if you are familiar with Docker, check out Python in Docker and Docker on Windows.

- **Mac OS Installation**

If you are on Mac OS X, it is recommended that you use Home brew to install Python 3. It is a great package installer for Mac OS X and it is really easy to use. If you don't have

Homebrew, you can install it using the following command:

$ ruby -e "$ (curl -fsSL

https://raw.githubusercontent.com/Homebrew/install/master/install)"

We can update the package manager with the command below:

$ brew update

Now run the following command to install Python3 on your system:

$ brew install python3

IV. Setting up PATH

Programs and other executable files should be in a variety of directories to provide the operating systems with a search path listing the directories that the OS is searching for.

The route is stored as a named string maintained by the operating system in an environment variable. The command Shell and other programs are supplied with this variable.

For Unix and Windows, the path attribute is called PATH (Unix is case-sensitive; Windows does not).

The installer manages the track information in Mac OS. You must add a Python directory to your path to invoke the Python interpreter from any specific directory.

- Setting Path at Unix/Linux

  To add the Python directory to the path for a particular session in Unix −

  - In the csh shell − type setenv PATH "$PATH:/usr/local/bin/python" and press Enter.
  - In the bash shell (Linux) − type export `ATH="$PATH:/usr/local/bin/python"` and press Enter.
  - In the sh or ksh shell − type `PATH="$PATH:/usr/local/bin/python"` and press Enter.

  Note: /usr/local/bin/python is the path of the Python directory.

- Setting Path at Windows

  To add the Python directory to the path for a particular session in Windows −

  At the command prompt −

  type path %path%;C:\Python and press Enter.

  Note : C:\Python is the path of the Python directory.

V. Running Python with GUI

If you have a GUI on your machine to support Python, you can run Python from a GUI environment as well.

- **Unix:** IDLE is the very first Unix IDE for Python
- **Windows:** Python Win is the first Windows interface for Python and is an IDE with a GUI.
- **Macintosh:** The Macintosh version of Python along with the IDLE IDE is available from the main website, downloadable as either Mac Binary or Bin Hex'd files.

You should take support from your system manager if you are unable to set up the framework properly. Ensure sure that the Python system is properly designed and runs well.

Another site called Anaconda can be used. We are also open. It includes hundreds of common data science packages, the conda package and the Windows, Linux and MacOS virtual environment manager. You can access it from the https:/www.anaconda.com/download/ link using your operating system.

In this book all the programs use Python 3.6 on MS Windows 10.

VI. Getting Started with python

(a) First Step:

Now in Python we're going to see how to run a standard 'Hello World.' You'll learn how Python programs are written, saved and run.

There are two methods of using the interactive interpreter prompt or using a source file with Python to run the software. let's now see how these two approaches are used.

By the Interpreter Prompt:

Open the terminal in your operating system and then open the Python prompt with the python3 key. The [enter] key will be pressed and open.

You will see >> when you start typing stuff once you started Python. This is known as the python interpreter prompt.

At the Python interpreter prompt, type:

```
>>print ("Hello World")
```

Then press the [Enter] key and as an output it will give-> Hello World.

Here is an example how screen will look in Fig 2.1

Fig. 2.1 Python Example in IDE

Quitting the Interpreter Prompt

You can exit the interpreter by pressing [ctrl + d] or entering exit) (if you're using the GNU/Linux or OS X shell (note: remember to use parentheses,))), (and then the key [enter].

On command prompt/ interpreter for Windows, press [ctrl+z] and then press [enter].

Using an Editor

A tool for writing and editing code is a code editor. Which are generally easy to know and can be fantastic. However, when your software is bigger, you will have to check and debug your code.

Your code is a lot better understanding than a text editor in the IDE (integrated developing environment). It usually includes features including construct automation, application layout, testing and debugging. It will improve your research considerably. The downside is that IDEs are difficult to use.

There are so many Editors like python IDE, Atom, PyCharm, Visual Studio code etc. These can be used for programming purpose and for development of various projects.

(b) Basics:

It's not enough just to print hello world, is it? How would like to do something else -like to make some changes, use them and get something out of them. In Python, by using constants and variables it can be done, and in this chapter, several other concepts will be discussed.

COMMENTS

Comments are a text to the right of the # symbol and are often useful for technology readers.

For example:

```
>>print ('hello world') # Note that print is a function
```

or:

```
# Note that print is a function
>>print ('hello world')
```

Use as many helpful comments as possible to clarify important decisions in your program to clarify important facts, explain problems you are trying to solve, explain problems you are trying to resolve, respectively.

Code shows you how, comments explains why.

LITERAL CONSTANTS

A number like 5, 1.23, or a string like 'This is a string' or 'It's a string,' an example of literal constant.

It is referred to as literal-you are simply using its meaning. The number 2 is always and is nothing else-it is a constant, because its value can't be modified. All of these are also called literal constants.

NUMBERS
Two types of numbers which are Integers and floats are commonly used.

One example is 2 that is just a number.

Examples of float numbers are 3.23 and 52.3E-4 (or shorter floats). The E notation displays 10 competencies. 52.3E-4 is in this case 52.3 * 10 ^ -4^.

STRINGS
A string is a character set. In essence, strings are just a bunch of words.

Nearly every Python program you write will have strings used, so be very careful about the following section.

SINGLE QUOTE

You can specify strings using single quotes such as 'Quote me on this'.

All white space i.e., spaces and tabs, within the quotes, are preserved as-is.

DOUBLE QUOTE

Strings in double quotes work exactly the same way as strings in single quotes. An example is "What's your name?"

TRIPLE QUOTE

You can specify multi-line strings using triple quotes - ("""" or '''). You can use single quotes and double quotes freely within the triple quotes.

An example is:

'''This is a multi-line string. This is the first line.

This is the second line.

"What's your name?," I asked.

He said "Bond, James Bond."
'''

Strings are IMMUTABLE: This means you can't alter a string once you've created it. And if it may seem a negative thing, it isn't really. We can see how the different services that we see afterwards are not limited by this.

VARIABLES:
The only way to store and manipulate data would soon be to be repetitive with only literal constants. Variables enter the view here.

Variables are just what the name means-their values can differ, i.e., something can be stored with a variable. Variables are just part of the memory of your machine in which information is stored. You need a way to control these variables and then call them, as opposed to literal constants.

IDENTIFIERS
Variables are examples of identifiers. Identifiers are names given to identify something.

There are some rules you have to follow for naming identifiers:

- The first character of the identifier must be a letter of the alphabet (uppercase ASCII or lowercase ASCII or Unicode character) or an underscore (_).
- The rest of the identifier name can consist of letters (uppercase ASCII or lowercase ASCII or Unicode character), underscores (_) or digits (0-9).
- Identifier names are case-sensitive. For example, myname and myName are not the same. Note the lowercase n in the former and the uppercase N in the latter.
- Examples of valid identifier names are i, name_2_3. Examples of invalid identifier names are 2things, this is spaced out, my-name and >a1b2_c3.

DATA TYPES

Different types of variables can hold values called data types. The key forms, which we have described, are numbers and strings. We will discuss later chapters how our own classes are to be formed.

Python refers to anything used as an entity in a program. It is generically said. They say "the object," rather than "the something,"

Python is highly object-oriented, in the sense that all is an entity, which includes numbers, strings and functions.

We will now see how variables with concrete constants can be used. Run the program and save the following example.

How to Write a program in python:
Steps to follow:

- Open editor/Interpretor
- Type the program code as in the example
- Save it as a file with a name with a [DOT]py or ".py" Extension
- Run the interpretor with an command *python filename.py* to run the program or you can use ru button on an editor to run the program.

Note: Filename can be anything but it should have an .py extension

Example: Type and run the following program

```
>>> # Filename : var.py
i = 5
print (i)
i = i + 1
print (i)
s = '''This is a multi-line string.
This is the second line
Welcome to artificial intelligence.'''
print (s)
```

```
Output:
5
6
This is a multi-line string.
This is the second line
Welcome to artificial Intelligence.
```

How It Works

Here's how this program works. First, we assign the literal constant value 5 to the variable i using the assignment operator (=). This line is called a statement because it states that something should be done and in this case, we connect the variable name i to the value 5. Next, we print the value of i using the print statement which, unsurprisingly, just prints the value of the variable to the screen.

Then we add 1 to the value stored in i and store it back. We then print it and expectedly, we get the value 6.

Similarly, we assign the literal string to the variable s and then print it.

INDENTATION

Whitespace in Python is significant. White space is actually necessary at the start of the line. That is what is known as indentation. At the beginning of the logical line, the leading whitespace, or tabs, is used to decide the indentation level of the logical line that is used in effect to define the aggregation of statements.

That implies all statements that are in general will have the same indentation. Both these statements are referred to as a block. Throughout the following pages, we can see examples of how blocks matter.

One thing you should know is that misplacement will lead to errors in the program. For instance:

```
>>> i = 5
# Error below! Notice a single space at the start of the line
 print ('Value is', i)
print ('I repeat, the value is', i)


You get the following error while running this:
File "whitespace.py", line 3
 print ('Value is', i)
 ^

Indentation Error: unexpected indent
```

Note at the beginning of the second line that there is a single space. Python's error means that the program's syntax is not accurate, and that the program has not been written correctly. It ensures for you that new blocks of statements cannot be started randomly (except for the existing principal block you've used all along).

OPERATORS & EXPRESSIONS

The majority of statements you write (logical lines) will contain words. An expression 2 + 3 is a simple example. An expression can be divided into operators and operands.

Operators are functions that can do anything, symbols like + or special keys are symbolic of them. Operators need other data and those data are referred to as operands. 2 and 3 are technicians in this case

We will discuss the operators and their use briefly.

Note that the expressions in the examples can be tested interactively with the interpreter. For instance, the interactive Python interpreter prompt is used to check expression 7 + 5:

```
>>>7 + 5
12
>>>3 * 8
24
>>>
```

Here is a short rundown of the available operators:

- + (plus)

  Adds two objects

  + 5 gives 8. 'a' + 'b' gives 'ab'

- - (minus)

  Gives the subtraction of one number from the other; if the first operand is absent it is assumed to be zero.

  -5.2 gives a negative number and 50 - 24 gives 26

- * (multiply)

  Gives the multiplication of the two numbers or returns the string repeated that many times.

  * 3 gives 6. 'la' * 3 gives 'lalala'

- ** (power)

  Returns x to the power of y

  ** 4 gives 81 (i.e., 3 * 3 * 3 * 3)

- / (divide)

  Divide x by y

  3 gives 4.333333333333333

- // (divide and floor)

  Divide x by y and round the answer down to the nearest integer value. Note that if one of the values is a float, you'll get back a float.

  13 // 3 gives 4

-13 // 3 gives -5

9//1.81 gives 4.0

- % (modulo)

  Returns the remainder of the division

  3 gives 1. -25.5 % 2.25 gives 1.5.

- << (left shift)

  Shifts the bits of the number to the left by the number of bits specified. (Each number is represented in memory by bits or binary digits i.e., 0 and 1)

  2 << 2 gives 8. 2 is represented by 10 in bits.

  Left shifting by 2 bits gives 1000 which represents the decimal 8.

- >> (right shift)

  Shifts the bits of the number to the right by the number of bits specified.

  11 >> 1 gives 5.

  11 is represented in bits by 1011 which when right shifted by 1 bit gives which is the decimal 5.

- & (bit-wise AND)

  Bit-wise AND of the numbers

  5 & 3 gives 1.

- | (bit-wise OR)

  Bitwise OR of the numbers

  5 | 3 gives 7

- ^ (bit-wise XOR)

  Bitwise XOR of the numbers

  ^ 3 gives 6

- ~ (bit-wise invert)

  The bit-wise inversion of x is - (x+1)

  ~5 gives -6.

- < (less than)

  Returns whether x is less than y. All comparison operators return True or False. Note the capitalization of these names.

  5 < 3 gives False and 3 < 5 gives True.

  Comparisons can be chained arbitrarily: 3 < 5 < 7 gives True.

- (greater than)

  Returns whether x is greater than y

  5 > 3 returns True. If both operands are numbers, they are first converted to a common type. Otherwise, it always returns False.

<= (less than or equal to)

Returns whether x is less than or equal to y

x = 3; y = 6; x <= y returns True

>= (greater than or equal to)

Returns whether x is greater than or equal to y

x = 4; y = 3; x >= 3 returns True

- == (equal to)

  Compares if the objects are equal

  x = 2; y = 2; x == y returns True

  x = 'str'; y = 'stR'; x == y returns False

  x = 'str'; y = 'str'; x == y returns True

- != (not equal to)

  Compares if the objects are not equal

  x = 2; y = 3; x != y returns True

- not (boolean NOT)

  If x is True, it returns False. If x is False, it returns True.

  x = True; not x returns False.

- and (boolean AND)

  x and y return False if x is False, else it returns evaluation of y

  x = False; y = True; x and y return False since x is False. In this case, Python will not evaluate y since it knows that the left-hand side of the 'and' expression is

  False which implies that the whole expression will be False irrespective of the other values. This is called short-circuit evaluation.

- or (boolean OR)

  If x is True, it returns True, else it returns evaluation of y

  x = True; y = False; x or y returns True. Short-circuit evaluation applies here as well

CONTROL FLOW

In the programs that we've seen until now, Python's exact top-down statements also were faithfully implemented. How if the rhythm of how it functions was to change? Do you, for example, like printing 'Good morning' or 'Good evening' depending on the daytime, want the system to take some decisions and do certain things according to the various situations?

As you might have imagined, the control flow statements are used to accomplish this. Within Python-if for and While-there are three statements of control flow.

- If Statement

  If the condition is true, we run a block (known as the if-block), otherwise we must process a block (known as the other block). If the condition is true, we must do that. This will be done. The other element is optional.

Example (save as if.py):

```
>>>number = 77
    guess = int (input ('Enter an integer : '))
    if guess == number:
    # New block starts here
        print ('Congratulations, you guessed it.')
        print (' (but you do not win any prizes!)')
    # New block ends here
elif guess < number:
    # Another block, in this we can use if for n number of times.
        print ('No, it is a little higher than that')
         # You can do whatever you want in a block...
        else:
print ('No, it is a little lower than that')
        # you must have guessed > number to reach here
print ('Done')
        # This last statement is always executed,
        # after the if statement is executed.:
```

OUTPUT:

```
$ python if.py
Enter an integer: 22
No, it is a little higher than that
Done
```

We take guesses from the user in this program and check that it is the number we contain.

We have specified the number of the variable to any integer, say 23. We then use the input function to guess the user. Which are Reusable pieces of programs are just featuring. We provide a string to the integrated input feature to print on the screen and wait for user input. The input) (function returns what we have entered as a string after we type something and click the kbd:[enter] key. This string is translated into an integer with int and stored in the conjecture variable. In general, the int is a class, but all that you have to learn right now is that you can use it to convert a string to an integer.

Next, we equate the user's guess to our selected number. by printing a success message if they are equal. Note, to say Python what statements to which block are used in indentation stages. In Python, indentation is therefore so necessary.

Note how the statement if at the bottom contains a column – we inform Python that it follows a block of statements.

Then we test if the assumption is less than the number, and if so, we will tell the consumer a bit higher than the assumptions. The elif clause that have been used here simply incorporates two similar if not other claims into a single if-elif else declaration. It improves the system and reduces the amount of indentation required.

At the end of the logic section, elif and other statements must always have a colon followed by a series of statements (which of course is properly indented).

In the if-block of an if statement and so on, you can have another if statement-it's called a nested if statement.

Remember that the elif and else parts are optional. A minimal valid if statement is:

```
>>>if True:
    print ('Yes, it is true')
```

Upon Python completes the complete if declaration along with the corresponding elif and other clauses, it moves on to the next statement in the if declaration sequence. The next statement is the print ('Done) 'statement. This is a big block (where execution of the program starts). Then Python sees and finishes the end of the plan.

- For Statement

The statement for .. in is another blowing statement, which passes through each element in a sequence through a sequence of items. In the following sections we will see more about series. Now you just have to know that a series is just a sequence of objects.

Example (save as for.py):

```
for i in range (1, 5):
print (i)
else:
print ('The for loop is over')
```

Output:

```
$ python for.py
1
2
3
4
The for loop is over
```

We print a sequence of numbers in this program. We use the built-in range function to produce this number series. All we do here is supply it with two numbers and return a series of numbers from the first to the second number. Range (1,5) for example offers the [1, 2, 3, 4] sequence. The range takes by default a move number of 1. The phase number is supplied if we supply a third number in the sequence. Range (1,5,2), for example, lists [1,3]. Notice that the range extends up to the second number i.e., the second number is not included.

Remember that, if you want a complete number set, for example, the list (range (5)) would result in [0,1, 2, 3, 4] list (s) is explained in the data structure chapter of the data structures class. The list (s) is then iterated for the loop-for the I in the field (1,5), the list (s) in [1, 2, 3, 4] is equal to the assignment of each number (s) in the series In this case, the value is only printed in the statement row.

Notice that the other segment is optional. When used, it is only done only after the loop has finished, unless a break statement is found. Note that for every sequence the.. in loop works. Here, we have a list of numbers created by the integrated range function, but we can generally use any type of object sequence!.

- While Statement

The while statement lets you execute a series of statements repeatedly, given that a condition is valid. An example of a so-called looping statement is a certain statement. An optional clause can be included in a certain sentence.

Example (save as while.py):

```
number = 23
running = True
while running:
guess = int (input ('Enter an integer : '))
            if guess == number:
print ('Congratulations, you guessed it.')
                        # this causes the while loop to stop
running = False
elif guess < number:
print ('No, it is a little higher than that.')
else:
print ('No, it is a little lower than that.')
else:
print ('The while loop is over.')
# Do anything else you want to do here
print ('Done')
```

Output:

```
$ python while.py
Enter an integer : 50
No, it is a little lower than that.
Enter an integer : 22
No, it is a little higher than that.
Enter an integer : 23
```

```
Congratulations, you guessed it.
The while loop is over.
Done
```

We are still playing the devination game in this program, but the benefit is that the user can only deviate until he thinks right, as we did in the last segment-there is no need to run it repeatedly. This shows the use of the While Statement properly.

We shift the input to and from the while loop and set the variable to True before the while loop. First, we test if the variable is true and then proceed to execute it during the process. After this block has been completed, the condition is tested again which is the running variable in this case. If it is valid, we will again run the While block, otherwise the optional other block will continue to be executed, so the following sentence will follow.

If the condition is incorrect the other block is executed, even though the condition is verified for the first time. If there's another clause for a certain loop, it will only work if you don't break out of the loop.

Truth and False are called Boolean forms and can be seen as being similar to 1 and 0.

THE BREAK STATEMENT

The break statement is used to break out of a loop statement i.e.,, avoid a loop declaration, even though it has not become false or the item sequence is not completely iterated.

A significant point is that every accompanying loop block is not carried out when you break off a for or during loop.

Example (save as break.py):

```
while True:
s = input ('Enter something : ')
    if s == 'quit':
break
    print ('Length of the string is', len (s))
print ('Done')
```

Output:

```
$ python break.py
Enter something : Programming is fun
Length of the string is 18
Enter something : When the work is done
Length of the string is 21
Enter something : if you wanna make your work also fun:
```

```
Length of the string is 37
Enter something : use Python!
Length of the string is 11
Enter something : quit
Done
```

THE CONTINUE STATEMENT

Using the continuation rule, Python will save the rest of the statements in the current loop block and start with the next iteration of the loop.

Example (save as continue.py):

```
while True:
        s = input ('Enter something : ')
        if s == 'quit':
        break
if len (s) < 3:
        print ('Too small')
        continue
    print ('Input is of sufficient length')
# Do other kinds of processing here...
```

Output:

```
$ python continue.py
Enter something : a
Too small
Enter something : 12
Too small
Enter something : abc
Input is of sufficient length
Enter something : quit
```

FUNCTIONS

Reusable program pieces are functions. You can assign a name to a block of statements, which you can use anywhere and any number of times to run the block using the defined name in your program. It is called function calling. Several integrated features like len and range have already been used.

Possibly the main building block for any untrivial applications (in any language of programming). Using the def keyword, functions are specified. After this keyword, a function name follows a couple of parentheses that may include the names of variables and the final column ending the line. The set of declarations that are part of this feature will follow next. An example shows that it's really simple:

Example (save as function1.py):

```
def say_hello ():
            # block belonging to the function
            print ('Welcome to agriculture')
            # End of function


say_hello () # call the function
say_hello () # call the function again
```

Output:

```
$ python function1.py
Welcome to agriculture
Welcome to agriculture
```

LOCAL VARIABLES

When you declare variables inside a function definition, they are not related in any way to other variables with the same names used outside the function - i.e., variable names are local to the function. This is called the scope of the variable. All variables have the scope of the block they are declared in starting from the point of definition of the name.

Example (save as function_local.py):

```
x = 50
def func (x):
        print ('x is', x)
        x = 2
        print ('Changed local x to', x)
func (x)
print ('x is still', x)
```

Output:

```
$ python function_local.py
x is 50
Changed local x to 2
x is still 50
```

THE GLOBAL STATEMENT

You must tell Python that a given name is not local, but global if you are to assign a meaning to a name at the top level of the program (i.e.,, not within any scope, such as the functions or classes). We use the global statement to do this. Without the global statement, a value can not be assigned to a variable specified outside of a function.

You may use values specified outside of the feature. This is not recommended and should be avoided because the user of the program does not know where the meaning of this variable is. It is amply clear from the global statement that the variable is specified in the external block.

Example (save as function_global.py):

```
x = 50
def func ():
        global x
        print ('x is', x)
        x = 2
        print ('Changed global x to', x)
func ()
print ('Value of x is', x)
```

Output:

```
$ python function_global.py
x is 50
Changed global x to 2
Value of x is 2
```

DEFAULT ARGUMENT VALUES

For certain features, if the user does not want to supply them with values, you can optionize for those parameters and use default values. The default argument values are used to do that. By adding parameter names to the parameter name of the assignment operator (=) specification, you can set the default parameter value for parameters.

Remember that the default statement is a constant value. More specifically-it is discussed in detail in later chapters-the default argument values must be immutable.

Example (save as function_default.py):

```
def say (message, times=1):
        print (message * times)
say ('Hello')
say ('World', 5)
```

Output:

```
$ python function_default.py
Hello
WorldWorldWorldWorldWorld
```

KEYWORD ARGUMENTS

The special syntax **kwargs in function definitions in python is used to pass a key worded, variable-length argument list. We use the name kwargs with the double star. The reason is because the double star allows us to pass through keyword arguments (and any number of them).

A keyword argument is where you provide a name to the variable as you pass it into the function.

One can think of the kwargs as being a dictionary that maps each keyword to the value that we pass alongside it. That is why when we iterate over the kwargs there doesn't seem to be any order in which they were printed out.

Example (save as function_keyword.py):

```
def func (a, b=5, c=10):
            print ('a is', a, 'and b is', b, 'and c is', c)
func (3, 7)
func (25, c=24)
func (c=50, a=100)
```

Output:

```
$ python function_keyword.py
a is 3 and b is 7 and c is 10
a is 25 and b is 5 and c is 24
a is 100 and b is 5 and c is 50
```

VARARGS PARAMETERS

Sometimes you might want to define a function that can take any number of parameters, i.e.,, variable number of arguments, this can be achieved by using the stars (save as function_varargs.py):

```
def total (a=5, *numbers, **phonebook):
print ('a', a)
#iterate through all the items in tuple
for single_item in numbers:
            print ('single_item', single_item)
            #iterate through all the items in dictionary
for first_part, second_part in phonebook.items ():
print (first_part, second_part)
total (10,1,2,3,Jack=1123,John=2231,Inge=1560)
```

Output:

```
$ python function_varargs.py
a 10
```

```
single_item 1
single_item 2
single_item 3
Inge 1560
John 2231
Jack 1123
```

RETURN STATEMENT

The return statement is used to return from a function i.e., break out of the function. We can optionally return a value from the function as well.

Example (save as function_return.py):

```
def maximum (x, y):
        if x > y:
                return x
        elif x == y:
                return 'The numbers are equal'
        else:
                return y
print (maximum (2, 3))
```

Output:

```
$ python function_return.py
3
```

DATA STRUCTURES

Essentially, data structures are just that-structures that can keep all data together. In other words, a list of similar information is processed. There are in the Python list, tuple, dictionary and set four built-in data structures. We'll see how each of them can be used and how they improve our life.

- **LIST:** A list is the data structure that comprises an ordered set of objects, i.e., a series of objects can be stored in a list. You can imagine this easily, if you can think of a shopping list with a list of things to buy except that each item is actually put separately on your shopping list, while Python lists commas.

    In square brackets the items collection should be included, so Python can understand that you define a collection. You can add, delete or check for the things in the list once you have built a list. we are able to add and remove items, a list is a mutable data type, i.e.,, it can be modified.

    Example (save as list.py):

```
# This is my shopping list
shoplist = ['apple', 'mango', 'carrot', 'banana']
print ('I have', len (shoplist), 'items to purchase.')
```

```
print ('These items are:', end=' ')
for item in shoplist:
    print (item, end=' ')
print ('\nI also have to buy rice.')
shoplist.append ('rice')
print ('My shopping list is now', shoplist)
print ('I will sort my list now')
shoplist.sort ()
print ('Sorted shopping list is', shoplist)
print ('The first item I will buy is', shoplist[0])
olditem = shoplist[0]
del shoplist[0]
print ('I bought the', olditem)
print ('My shopping list is now', shoplist)
```

Output:
```
$ python ds_using_list.py
I have 4 items to purchase.
These items are: apple mango carrot banana
I also have to buy rice.
My shopping list is now ['apple', 'mango', 'carrot', 'banana',
                         'rice']
I will sort my list now
Sorted shopping list is ['apple', 'banana', 'carrot', 'mango',
                         'rice']
The first item I will buy is apple
I bought the apple
My shopping list is now ['banana', 'carrot', 'mango', 'rice']
```

- **TUPLE:** Tuples are used to hold several objects together. Think of them as lists, but without the broad features that you get from the list class. One of the most important characteristics of tuples is their unchangeable nature, that is, you can't alter tuples.

  Tuples are identified in the optional parenthesis by indicating commas-split objects. Tuples are typically used where a statement or function specified by the user may assume safely that the set of values does not change (i.e., the multiple of the values used).

  Example (save as tuple.py):

```
# I would recommend always using parentheses
# to indicate start and end of tuple
# even though parentheses are optional.
```

```
# Explicit is better than implicit.
zoo = ('python', 'elephant', 'penguin')
print ('Number of animals in the zoo is', len (zoo))
new_zoo = 'monkey', 'camel', zoo
    # parentheses not required but are a good idea
print ('Number of cages in the new zoo is', len (new_zoo))
print ('All animals in new zoo are', new_zoo)
print ('Animals brought from old zoo are', new_zoo[2])
print ('Last animal brought from old zoo is', new_zoo[2][2])
print ('Number of animals in the new zoo is', len (new_zoo)
                              1+len (new_zoo[2]))
```

Output:

```
$ python ds_using_tuple.py
Number of animals in the zoo is 3
Number of cages in the new zoo is 3
All animals in new zoo are ('monkey', 'camel', ('python',
                      'elephant', 'penguin'))
Animals brought from old zoo are ('python', 'elephant',
                      'penguin')
Last animal brought from old zoo is penguin
Number of animals in the new zoo is 5
```

DICTIONARY

A dictionary is like an address-book in which you can find an individual's address or contact information only with the individual's name, that is, associating keys (name) with values (details). Please remember that the key must be special, such like if you have two people of the same name, you cannot find the correct keys.

You can only use immutable (such as strings) objects for dictionary keys, but for dictionary values you can use dynamic or moving objects. This simply means that only simple objects can be used for keys.

The dictionary defines keys and values pairs by notation

d = {key1: value1, key2: value2}.

Note that the key value pairs are separated by one colon and the pairs are divided by comas, all of which are contained in a pair of curly braces.

Note that key value pairs are not ordered in a dictionary in any way. If you want a specific order, you must sort it yourself before using it. The dictionaries you can use are dict type instances/objects.

Example (save as dict.py):

```
# 'ab' is short for 'a'ddress'b'ook
ab = {'Swaroop': 'swaroop@swaroopch.com','Larry': 'larry@wall.org',
     'Matsumoto': 'matz@ruby-lang.org',
     'Spammer': 'spammer@hotmail.com'
     }
print ("Swaroop's address is", ab['Swaroop'])
# Deleting a key-value pair
del ab['Spammer']
print ('\nThere are {} contacts in the address-book\n'.format
     (len (ab)))
for name, address in ab.items ():
print ('Contact {} at {}'.format (name, address))
# Adding a key-value pair
ab['Guido'] = 'guido@python.org'
if 'Guido' in ab:
        print ("\nGuido's address is", ab['Guido'])
```

Output:

```
$ python ds _ using _ dict.py
Swaroop's address is swaroop@swaroopch.com
There are 3 contacts in the address-book
Contact Swaroop at swaroop@swaroopch.com
Contact Matsumoto at matz@ruby-lang.org
Contact Larry at larry@wall.org
Guido's address is guido@python.org
```

- **SET:** Sets of simple items are unordered arrays. They are used where something is more important than the order or how much the presence of an item in a set is.

  You can use sets to check membership if it is a subset of another set, find the crossing between two sets, etc.

  For example (run it in python interpretor or in IDE):

```
>>>bri = set (['brazil', 'russia', 'india'])
>>>'india' in bri
True
>>>'usa' in bri
False
>>>bric = bri.copy ()
```

```
>>>bric.add ('china')
>>>bric.issuperset (bri)
True
>>>bri.remove ('russia')
>>>bri & bric # OR bri.intersection (bric)
{'brazil', 'india'}
```

OBJECT ORIENTED PROGRAMMING

We have built our program for all the programs which we have written so far in order to manipulate data, i.e., blocks of statements. It is known as the process-oriented programming task. Another way to organize the program is to combine data and features and bundle them in something that is called an entity. The programming style is called object orientated. You may typically use procedural programming, but you can use object-orientated programming techniques when writing large programs or if there is a problem which is better suited to this process.

The two key aspects of object-oriented programming are classes and objects. A class provides a different kind in which objects are class instances. An analogy is that you can have int variables that indicate that the integer saving variables are int class instances (objects).

Objects can store data using standard object variables. Fields are referred to as attribute belonging to an object or class. Objects may also provide functions through the use of class-owned functions. These functions are known as class methods. This terminology is important since it allows us to distinguish between independent and class or object-specific functions and variables. The fields and methods may collectively be referred to as the class attributes.

Fields are two-type-they can be categorized into any instance/object or they can be categorized by themselves. They are respectively referred to as variables of instance and class.

Using the keyword class is developed. The class fields and methods are shown in a series of indents.

The SELF: There is only a single particular difference to the class methods from ordinary functions-they require an additional first name to be applied to the start of the parameter list, but when you call a method, Python doesn't give this a meaning. This attribute refers to the object itself and is called self by default.

While it is highly advised that you use the self name for this parameter-it is undoubtedly conceivable that any other name is frowned upon. There are several benefits of using the common name-any user of your software can know it instantly, and even advanced IDEs.

You have to wonder how Python gives self meaning and why you should not give it meaning. This is made clear by an illustration. Say you have the MyClass class and a myObject class case. When you call a myobject.method (arg1,arg2) method this is translated by Python automatically into MyClass.method (myobject, arg1,arg2) –

that's about the whole thing. This is a unique self.

This also implies even if you have an argument-free process, there is always an argument-the "self."

CLASSES:

The simplest class possible is shown in the following example (save as oop_simplestclass.py).

```
class Person:
pass # An empty block
p = Person ()
print (p)
```

Output:

```
$ python oop_simplestclass.py
<_ _main_ _.Person instance at 0x10171f518>
```

Here we see the self in action. Notice that the say_hi method takes no parameters but still has the self in the function definition.

The __init__ Method:

In the Python classes there are several method names that have a specific relevance. The sense of the init process is now clear.

If an object of a class is instantiated (e.g., created), the init method is executed.

The method is useful for any initialization you want to create (i.e., pass initial values to your object). Please note the double focus at the start and the finish of the term.

Example (save as oop_init.py):

```
class Person:
        def _ _init_ _ (self, name):
        self.name = name
def say_hi (self):
print ('Hello, my name is', self.name)
p = Person ('Mahesh')
p.say_hi ()
# The previous 2 lines can also be written as
# Person ('Swaroop').say_hi ()
```

Output:

```
$ python oop_init.py
Hello, my name is Mahesh
```

Here the init approach is specified as the name of a parameter (including the popular self). We're just creating a new area named name here. Remember that there are two variables as both are named 'name.' Both are distinct. The dotted notation aut. name indicates that something named "name" is part of the object "self," and a local variable is the other name. There is no ambiguity as we specifically state the name to which we refer.

When you construct new instance p, of the person class, we use the name of the class and then the arguments in the parentheses: p = Person ('Mahesh').

We do not explicitly call the __init__ method. This is the special significance of this method.

Now, we are able to use the self.name field in our methods which is demonstrated in the say_hi method.

CLASS and OBJECT VARIABLES

The functionalities of classes and objects (i.e., methods) have been addressed already, and we can now hear about the data component. The data portion, i.e., fields, are just ordinary variables that are connected to class and object name spaces. It means that these names only refer to these classes and objects. That's why name spaces are named.

Two types of fields are classified according to whether the class or object is the property of the variables respectively

Class variables are shared-all instances of that class can access them. There is only 1 copy of the class variable, which will be used by all other instances if an object transitions to a class variable.

Every class object/instance is held by the object variables. In this case, each object has its own copy of the field, i.e., in a separate instance it is not shared and in no way is connected to the field by the same name. This will be easy to understand by an example (save as oop objvar.py):

```python
class Robot:
"""Represents a robot, with a name."""
        # A class variable, counting the number of robots
population = 0
def _ _init_ _ (self, name):
        """Initializes the data."""
        self.name = name
        print (" (Initializing {})".format (self.name))
        # When this person is created, the robot
        # adds to the population
        Robot.population += 1
def die (self):
        """I am dying."""
```

```
        print ("{} is being destroyed!".format (self.name))
        Robot.population -= 1
        if Robot.population == 0:
                print ("{} was the last one.".format (self.name))
        else:
                print ("There are still {:d} robots working.".format
                                                (Robot.population))


def say_hi (self):
        """Greeting by the robot.
        Yeah, they can do that."""
        print ("Greetings, my masters call me {}.".format
                                                (self.name))

        @classmethod
        def how_many (cls):
                """Prints the current population."""
                print ("We have {:d} robots.".format (cls.population))

droid1 = Robot ("R2-D2")
droid1.say_hi ()
Robot.how_many ()
droid2 = Robot ("C-3PO")
droid2.say_hi ()
Robot.how_many ()
print ("\nRobots can do some work here.\n")
print ("Robots have finished their work. So let's destroy them.")
droid1.die ()
droid2.die ()
Robot.how_many ()
```

Output:
```
$ python oop_objvar.py
(Initializing R2-D2)
Greetings, my masters call me R2-D2.
We have 1 robots.
(Initializing C-3PO)
Greetings, my masters call me C-3PO.
We have 2 robots.
```

```
Robots can do some work here.
Robots have finished their work. So let's destroy them.
R2-D2 is being destroyed!
There are still 1 robots working.
C-3PO is being destroyed!
C-3PO was the last one.
We have 0 robots.
```

It is a broad example of how class and object variables are to be illustrated. In this case, the population belongs to the Robot class and is thus a class attribute. The name variable is an object variable (the variable is allocated by itself).

We also refer to the population variable as Robot.population rather than Self.population. We use self.name notation in the object 's methods to refer to the object variable name. Notice this basic difference between the variables of class and object. Note also that the variable value would be obscured by an object of the same name as a value!

We should have been using self. class.population instead of Robot.population, because any object is referenced to its class through the self. class attribute.

How many is a function that does not belong to the object but rather to the class. This means that, depending on whether we need to learn which class we are in, we can describe it as either a dynamic method or a static one. We use the class form since we refer to a class attribute.   '

We have listed the many ways of using a decorator as a class tool.

Decorators can be thought of as shortcuts to a function called by the wrapper (e.g., a function that "wraps" around a function so it can do something before or after the internal operation). Thus, using the @classmethod decorator is the same as calling:

how many = classmethod (how many).

In this way, we increase the population number by 1 as we add another robot. Remember that self.name values are unique to each entity that shows the existence of the variables of the entity.

Remember that the variables and methods of the same object should be referred to using only the self. This is regarded as a reference attribute.

We also see in this software the use of class and process docstrings. We can use Robot. doc and the docstring method Robot.say hi. doc In the die method, we simply lower the Robot.population count by 1. In the die method, we may access a docstring class.

Each member of the class is public. One exception is that Python uses name-mangling to effectively render a private variable by means of names that use the double-underline prefix like private var.

The rule is that every variable to use within the class or object will start with the underscore, and all other names are public and can be used for all classes or objects.

This practice is followed. Recall that this is just a convention and not enforced by Python (except a prefix with double emphasis).

INHERITANCE

Reuse of code is one of the main benefits of object-orientated programming and one way to accomplish this is through the heritage mechanism. Inheritance can best be interpreted as a relationship between classes of type and subtype.

Suppose you want to write a program that tracks teachers and students in a university. They have some common features including name, age and address.

These have unique features, such as wage, courses and instructor fees and student charges.

You can build and process two independent classes for each type, but adding a common feature will add to these two independent classes. It's quickly becoming unmanageable.

This will be a great way to build a single class called Schoolmember and then have the instructor and student classes belonging to this class and then we can apply different features to those subtypes.

This strategy has several advantages. If in SchoolMember, we add/modify a function this is reflected automatically in the subtypes. For example, by simply adding it to the SchoolMember class, you can create a new ID card area for both teachers and students. But changes in the subtypes have no effect on other subtypes.

One benefit is that you can refer to a teacher or student object as an object SchoolMember that can be helpful in some circumstances such as the number of schoolmembers counting. This is called polymorphism where a sub-type can be substituted in any situation where a parent type is expected, i.e., the object can be treated as an instance of the parent class.

Notice that, in the case of separate classes, we reuse the class code and do not have to replicate it in various classes like we had to do.

In this case, the Schoolmember class is called the Main or Superclass class. The classes of teachers and students are called the derived grades.

For Example (save as inheritace.py):

```python
class SchoolMember:
        "'Represents any school member.'"
        def _ _init_ _ (self, name, age):
        self.name = name
        self.age = age
        print ('(Initialized SchoolMember: {})'.format (self.name))
def tell (self):
        "'Tell my details.'"
        print ('Name:'{}" Age:"{}"'.format (self.name, self.age),
                                                                end=" ")
```

```
class Teacher (SchoolMember):
"'Represents a teacher.'"
      def _ _ init _ _ (self, name, age, salary):
            SchoolMember. _ _ init _ _ (self, name, age)
            self.salary = salary
            print ('(Initialized Teacher: {})'.format (self.name))
      def tell (self):
            SchoolMember.tell (self)
            print ('Salary: "{:d}"'.format (self.salary))
class Student (SchoolMember):
      "'Represents a student.'"
      def _ _ init _ _ (self, name, age, marks):
            SchoolMember. _ _ init _ _ (self, name, age)
            self.marks = marks
            print ('(Initialized Student: {})'.format (self.name))
      def tell (self):
            SchoolMember.tell (self)
            print ('Marks: "{:d}"'.format (self.marks))


t = Teacher ('Mrs. Shrividya', 40, 30000)
s = Student ('Swaroop', 25, 75)
# prints a blank line
print ()
members = [t, s]
for member in members:
      # Works for both Teachers and Students
      member.tell ()
```

Output:
```
$ python oop _ subclass.py
(Initialized SchoolMember: Mrs. Shrividya)
(Initialized Teacher: Mrs. Shrividya)
(Initialized SchoolMember: Swaroop)
(Initialized Student: Swaroop)
Name:"Mrs. Shrividya" Age:"40" Salary: "30000"
Name:"Swaroop" Age:"25" Marks: "75"
```

FILES: By creating an object in the class of files and using the file's reading, reading or writing methods, you can open and use the file to read or write. It depends on the mode you have defined for file opening to read or write the file. Then you eventually call the close method when you finish the file to inform Python that we're finished using the script.

Example (save as file.py):

```python
poem = '''\
Programming is fun
When the work is done if you wanna make your work also fun: use
Python!'''
# Open for 'w'riting
f = open ('poem.txt', 'w')
# Write text to file
f.write (poem)
# Close the file
f.close ()
# If no mode is specified,
# 'r'ead mode is assumed by default
f = open ('poem.txt')
while True:
line = f.readline ()
# Zero length indicates EOF
if len (line) == 0:
break
                # The 'line' already has a newline
# at the end of each line
# since it is reading from a file.
print (line, end='')
# close the file
f.close ()
```

Output:

```
$ python3 io_using_file.py
Programming is fun
When the work is done
if you wanna make your work also fun:
use Python
```

Notice that by simply using open method we can create a new file object. Through using the built-in open feature and specifying a file name and the mode where we want to open the file, we open (or build it if it does not already exist). Mode can include the read ('r') mode, write ('w') or add ('a') mode. We may also indicate, in the text ('t') or in the binary ('b'), if you are reading, writing, or appending. In reality, several more modes are available and help (open) will give you more information.

Open) (detects the file by default as a 'text file and then opens it in' read mode.

In our example, we initially open/create the file in text mode and use the file object writing procedure for writing our variable string poem and then close the file.

First, for viewing, we open the same file again. We do not have to specify a mode because the default mode is 'read text file.' We read the readline method in a loop in each line of the file. The entire line with the newline character at the end of the line is restored. When we return a blank string, it means we are at the end of the file and break out of the loop.

In the end, we finally close the file.

We can see from our readline output that this program has indeed written to and read from our new poem.txt file.

Summary

As we have discussed the important basics of python and every concept of python programming language is discussed with suitable examples. Python will be used in further chapter implementation parts and for making real time projects which can be used in real time for research and academic studies. In the implementation of python for further branches of artificial Intelligence there are several Libraries and frameworks are involved which will be discussed from topic to topic with its installation and uses with different algorithms.

Chapter 3

Machine Learning

As we know there are different branches of artificial Intelligence and machine Learning of one of them which has become a most advance and widely used branch of artificial Intelligence in every field of this generation whether it may be either electronics, civil, mechanics, Bioengineering, medical or agriculture etc., In this chapter we will study about machine learning and its types with the most common algorithms of machine learning which are being mostly used in different scenarios and their implementation using python language has been discussed in the implementation part of this book. Here we will also see where machine learning can be used.

I. Introduction

Machine learning is a subset of Artificial Intelligence and one of the most important part of artificial Intelligence. It is a field of Computer science which is totally differs from traditional computational approaches. In traditional approaches, a set of instruction is explicitly provided to the algorithm to solve any problem. Whereas in Machine Learning is a domain of an AI which uses a statistical analysis and can enables a system to always learn from its data without any external inputs in order to give an essential output. Due to this machine learning makes computers in advance to make model which provide decision based on input data.

Machine Learning involves a lot of algorithms which usually based on their requirements and functionalities to use where and when. To make a machine learning model we always require the data on which model can be made. Machine learning uses the algorithm which continuously learns from the data available as training data. A machine learning model is generated after the training of algorithm from the data. After the training completes, when trained model is provided with other real-life data then you can get an essential outcome. For example, in figure 3.1 Machine learning Model we use a predictive algorithm and make a model by providing it with some data, and later on we can receive prediction's based on real data as input.

In any field, a user of technology has always been benefited from Machine learning. In artificial intelligence as Machine learning has been always been the continuously developing field and addition to this there always been consideration which has to be kept in mind while learning machine learning algorithms and their methodologies or their analysis into different field.

ITERATIVE LEARNING FROM DATA: As machine learning models always get trained from different datasets before deployed for use. Some of the model which are online and can get adapted to new data continuously. On other side, some machine learning model are offline which get trained and once deployed somewhere cannot be change and they always give the output based on those. The online models in machine learning performs iterative learning as they get new data continuously in which algorithms creates patterns and associates it with data elements present in the data. After a model gets trained it can be utilized for real time to learn from data and give us meaningful outcomes.

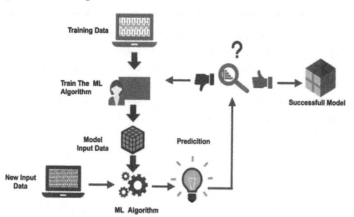

Fig. 3.1 Machine Learning Model

II. Approach to Machine Learning

When we talk about several type of problems to be solve by machine learning which involves different type of algorithm. In the same way we have different techniques in machine learning which can be utilized in different scenarios. For different problem statements we use different machine learning techniques which can be based on type and volume of data. Machine learning has different categories which are discussed here:

1. Supervised Learning

It is the most commonly and widely used Machine Learning, Model building is always depending on data available and type of output we require. As in supervised learning, the models are given with the already classified and labeled data for training purpose or to learn and later on by computing model with other actual data, the results are compared with the assumed or expected trained one's. If certain mismatch or error is found model is again trained with more data. As supervised learning always used to predict based on the patterns found into the data or on unlabeled data given to the model.

In other words, supervised learning can said to be a guided learning or teacher's learning where we have an already labelled dataset which acts as teacher to the Machine learning model to learn. If the model is trained, it will begin to predict or determine if new data are given. The details can be understood:

Suppose we have input variables, x and output variable, y, to learn how the mapping function can be learned from input to output,

$$\text{such as: } Y = f(x)$$

Now, the primary objective is to approximate the mapping function in a way that the output variable (Y) can be predicted for such data if we have new input (x).

Two types of problems can be classified into primarily Supervised learning:

Classification: A classification query is named when we have categorized performance such as "black," "teaching," "not teaching," etc.

Regression: A regression problem is named if the real output value, like "width," "kilogram," etc., is available.

Examples of supervised machine learning algorithms are the decision tree, random forest, knn, logistic regression.

A common application of supervised learning is to use historical data to statistically predict future events. You can either forecast upcoming movements using historical stock market knowledge or use spam mails to filter. Tagged plants or crops images in agriculture can be used as input data in supervised training for classifying untagged images of plants and crops.

2. **Unsupervised Learning:**

Unsupervised learning is best if the question involves a large amount of unlabeled data. Knowing the meaning behind these data includes algorithms that can be understood by classifying the data on the basis of the patterns or clusters found. The supervised learning thus carries out an iterative data processing process without human intervention.

The Algorithms in Unsupervised learning segment data into outstanding groups (clusters) or feature groups. Unlabeled data determines the values of the parameter and the data classification. This method basically applies labels to the information so that it is becomes supervised and model gets trained. Unsupervised learning will decide whether a large amount of data is available. The developer does not know the context of the data to analyze in this case, so at this point, the labeling is not possible. That's Why Unsupervised learning can therefore be used as the first step before data are moved to a regulated learning process. The details can be understood:

Suppose we have an input variable x, and no output variables are available as in supervised learning algorithms. Simply put, we may conclude that there is no right answer and no guidance instructor in unsupervised learning. Algorithms allow users to identify fascinating data trends.

The following two forms of problem can be divided into unsupervised learning problems:

Clustering: We must discover the underlying groupings of the data in problems of clustering. For instance, consumer grouping by shopping behavior.

Association: An association problem is named since these problems require that rules be discovered that explain many of our data for instance, to find the clients that buy both x and y.

In unsupervised learning algorithms can help understand large amounts of new, unlabeled content. These algorithms search for data patterns similar to supervised learning (see the preceding Section). The difference is that data are not already known.

For example, The Unsupervised machine learning algorithms are like K-means for clustering, Apriori algorithms for association.

In agriculture field, gathering vast quantities of data on a particular field will, for example, help farmers gain insight into trends and link them to crop harvesting and increase in production of it. All the data points linked to a condition like health of plants will take too long to classify. The unregulated learning strategy will also help to evaluate outcomes faster than a supervised approach to learning.

3. **Reinforcement Learning**:

Reinforcement learning is a model of behavioral learning. Feedback from the data analysis is received in order to guide the user in the best result. Increased learning differs from others because the system does not use a sample data set to train. Instead, the system learns by testing and error. A series of successful choices will therefore "reinforce" the process as the problem is best resolved.

Figure Agent→system

In the Figure which reflects the general scenario of improving learning. In comparison, the agent receives no designated data here, unlike the supervised learning scenario considered in the previous section. It gathers information instead by communicating with the world, via a series of acts. The agent receives two kinds of information in response to an event: his or her state of play in the world and the real-life scenarios, which is unique to the task and its respective purpose. The goal is for the Agent to maximize its reward and, in order to achieve that objective, to agree on the best course of action or policy.

The data he collects from the system, however, is just the immediate reward for the action he has taken. There is no input from the system on future or long-term incentives. A significant element of enhancement is to take account of late bonuses or penalties. The agent faces the dilemma between exploring unknown states and actions to gain more information about the System and the rewards and using the already accumulated information in order to maximize its reparation. This is known as the trade-off between exploration and development inherent in Reinforcement learning.

Understand that in several of the preceding chapters there are many variations between the reinforced learning scenario and the supervised learning. In comparison to supervised learning, there is no fixed distribution of the instances in reinforcement learning, which defines a distribution over observations. It is the choice of policy. In reality, minor policy changes can drastically affect rewards. In addition, the System cannot be set in general and can differ due to the chosen behavior by the agent. This may be more practical than normal supervised learning for certain learning problems.

There are two key settings: one in which the ambient model is known to the agent, which reduces its goal of optimizing the amount of money to a planning problem; and the other in which the ambient model is unknown and where the agent is faced with a learning question. In the above case, the agent must learn from the state and recompense data obtained to acquire relevant information and decide the best strategy.

Reinforcement Learning is not specifically supervised Learning as not strictly reliant on collection of data "supervised" (or labeled) (training collection). In reality, the response of actions taken can be controlled and calculated against a concept of "reward." However, it is also not unsupervised learning because we know the desired reward in advance when we form our "learner."

III. Supervised Vs Unsupervised Vs Reinforcement

Finally, as all are well aware of supervised, unsupervised and Reinforcement learning, look at the gap between these learning's.

To sum up, supervised learning happens when a model learns with guidance from an defined data set. And unsupervised learning is where the computer is trained without any guidance on the basis of unlabeled data. Reinforcement learning happens when a computer or an agent communicates, carries out actions and learns using a test-and-error process.

Criteria	Supervised Learning	Unsupervised learning	Reinforcement Learning
Definition	The Model learns using labelled or classified data	The model is trained with unclassified or unlabeled data	An agent interact with the model by taking action & learning from the errors and rewards.
Types of Problem	Regression & Classification	Association & Clustering	Reward based
Type of data	Labeled Data	Unlabeled data	No predefined data
Training	External Supervision	No supervision	No supervision
Approach	Maps the labeled or classified inputs to the known outputs	Recognizes the pattern in unclassified data and discovers the output	It follows the trail and error method

IV. Most Common Machine Learning Algorithms

- **Linear Regression:** This is one of the most popular statistics and machine learning algorithms.

   Linear regression is essentially a model of a linearity that assumes a linear relationship between the input variables x and y. In other words, you can calculate the input variables x from a linear combination. It can be defined by a best line to create a relationship between the variables. Linear Regression is generally of two types which are

   - **Simple Linear Regression:** A linear regression algorithm is considered a simple linear regression because there is just one independent variable.

- *Multiple Linear Regression:* If it has more than one independent variable, a linear regression algorithm is considered a multiple linear regression.

  Linear regression is used mainly for estimating true values based on continuous variable (s). For example, a full day sale of a shop can be calculated by a linear regression based on real values.

- **Logistic Regression:** It is also known as logit regression. It is a classification algorithm. Logistic regression is mainly a classification algorithm used to estimate discrete values like 0 or 1, true or false or yes or no based on a particular collection of separate variables. In theory, it predicts that its output will be between 0 and 1.

- **Decision Tree:** It is a supervised learning algorithm that is used mainly for problem classification. This is basically a graded partition based on the independent variables as a recursive partitive. Decision tree has rooted tree nodes. The Rooted Tree is a directed tree with a root node. Root has no input edges and all the other nodes have an input edge. Such nodes are referred to as leaves or nodes of judgment. Consider, for example, the following Fig 3.2 decision tree to test whether or not an individual is suitable for the job.

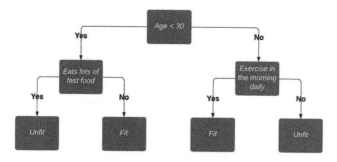

Fig. .3.2 Decision Tree

- **Support Vector Machine:** It is used for problems of classification and regression. However, it is primarily used for problems of classification. SVM's key principle is to assign each data element in the n-dimensional space as the value of the individual feature. Here n are the characteristics we have. Then a simple graphic representation to grasp the SVM definition from the Fig 3.3 SVM.

Fig. 3.3 SVM

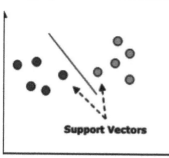

Within the above Fig 3.3 SVM, we have two features hence we first need to map such two variables within two-dimensional space where each point has two co-ordinates, called support vectors. The line divides the data into two separate classified categories..This line is the classifier in SVM

- **Naïve Bayes:** It's also a technique of classification. Bayes theorem is a principle for constructing classifiers behind this classification technique. The predictors are expected to be independent. In simple terms, the inclusion in a class of a particular characteristic is not connected to that of any other characteristic. The following is the Bayes theorem equation:

$$P(A|B) = P(B|A)\ P(A)/P(B)$$

  The model of Naïve Bayes is simple to develop and especially useful for large data sets

- **K-Nearest Neighbors (KNN):** It is used to identify problems and to rectify them. It is used commonly to solve problems with classification. The key principle of this algorithm is that it stores all the existing cases and by voting plurality of its neighbors classifies new cases. The case is then assigned to the class which, according to a distance function, is the most common among his nearest K neighbors. The gap can be from Minkowski, Euclidean, and Hamming. Consider KNN as follows:

  - Computationally KNN are expensive than other algorithms used for classification problems.
  - The normalization of variables needed otherwise higher range variables can bias it.
  - At KNN, we will work on a stage such as noise reduction, pre-processing.

- **K-Means Clustering:** This is used to resolve the problem of clusters, as the name implies. This is a kind of unsupervised Learning. The principal principle of the K-Means algorithm is to assign the data set across different clusters. Follow these steps to build K-means clusters:

  - For each cluster called centroids, K-means picks up the k number of the points.
  - Every data point now forms a cluster, i.e., k clusters, with the nearest centroids
  - Now, the centers, based on the current members of each cluster, will be identified.
  - These measures need to be repeated before convergence takes place.

- **Random Forest:** This is a supervised algorithm of classification. The random forest algorithm has the advantage that it could be used for problems of classification as well as regression. This is essentially the set of decision-making trees (forests) or the decision-making trees ensemble. The basic principle of the random forest is that each trees are graded and that the forest selects the best grades. The advantages of Random Forest algorithm are as follows:

  - For classification and regression functions, the random forest classifier may be used.
  - The missing values can be managed.
  - Even if we have more trees in the forest, this will not over suit the pattern.

V. Applications

- Retailers

  Machine Learning in agriculture is used by the seed retailers to turn data into better crop production. Through their use by pest control firms, the growing bacteria, bugs and vermin can be detected.

- Agriculture Bots

  The majority of companies are now preparing and building robots for the main agricultural mission. This involves seed processing and work more efficiently than manpower. This is the latest example of agricultural machine learning.

- Species Breeding

  Species selection is a repetitive searching method for specific genes, which decide the effectiveness of water and the use of nutrients, climate tolerance, disease resistance, nutrient content or taste. In the analysis of crop production in different climates and developing new technologies, machine-learning in particular, profound learning algorithms requires decades of field data. Based on this data, a statistical model can be developed that predicts which genes are most likely to contribute to a plant's gain.

- Species Recognition

  The conventional approach of the human person for plant classification is comparing the color and shape of the leaves, but by using Machine Leaning's statistical analysis can provide quicker and more detailed results by analyzing the morphology of the leaves and provides more details about the leaf properties.

- Yield Production

  Yield forecasting is one of the topics of steep-precision agriculture, as it defines yield mapping and prediction, crop demand matching and crop management. state-of - the-art techniques have gone far beyond simple prediction based through historical data, but have involved machine learning techniques to provide knowledge concerning crops, weather and economic circumstances and extensive multidimensional analysis to optimize the return for farmers and the public.

- Disease Detection

  Machine Learning can be used to detect Plant infections which are typically caused by plague, insects, diseases and, unless managed on time, they reduce productivity on a large scale. Concerning the area grown in acres, the cultivators are tedious to track the crops daily. By implementing Machine Learning here provides the solution to monitor the cultivated area consistently and provides automatic detection of diseases with remote sensing images.

- Crop Quality

  Data output can be adequately measured and defined to lift product price and reduce waste. Contrary to human experts, machines may use apparently irrelevant data and interfaces to discover and identify new qualities that play a role in the overall quality of the crops.

- Weed Detection

  Weeds are the major threats to crop production, apart from diseases. The main issue in the war against weeds is that they are hard to distinguish and discriminate against crops. Computer vision and ML algorithms are able to enhance weed identification and discrimination at low cost without environmental issues or side effects. The technologies are going to push robots to kill weeds and reduce the need for herbicides in future.

- Soil Management

  Soil is a highly diverse natural resource with complex processes and unclear mechanisms for agricultural specialists. Its own temperature can give insights into the impact on territorial returns of climate change. The algorithms of machine learning study the process of evaporation, soil humidity and temperature to understand ecosystem dynamics and the effects of agriculture.

- Water Management

  The hydrological, climatological and agricultural balance of water resources in agriculture has a significance. Using more efficiently irrigated systems and a forecast of a daily dewpoint temperature, the most advanced ML applications so far are connected with an estimation of daily, weekly, or monthly stomata, which helps to classify predicted weathers and to estimation the evaporation.

- Livestock Production

  Machine learning offers accurate prediction and calculation of farming parameters, in accordance with crop management, to maximize the economic efficiency of farming systems such as livestock production. In order to allow farmers to change diets and conditions, for example, weight forecasting system will be estimating future weight 150 days before the day they slaughter.

Summary

As Machine learning has been widely used in several field and it has become a most important branch of artificial Intelligence. As we have seen there can be a lot of used of Machine Learning in the field of agriculture which can help the farmers to increase their production of crops and save them from pests and other insects while harvesting in the field. So, we can see that how machine learning can be useful in the agricultural field.

Chapter 4

Deep Learning

Deep learning emerged as a serious contender in the field from a 10-year explosive computational development. Deep learning is therefore a special form of machine learning, the algorithms of which are based on the structure and function of the human brain.

Machine Learning VS Deep Learning

Deep learning is arguably the most efficient form of machine learning. This is so important because they know the best way to solve the problem and how to fix everything. The following is a description of deep learning and machine Learning:

- **Data Dependency:** The first difference point is based on the DL and ML output as the data scale increases. Deep learning algorithms are very good when the data is large.
- **Machine Dependency:** High-end machines allow deep learning algorithms to function perfectly. Machine learning algorithms, on the other hand, can also operate on low-end computers.
- **Machine dependency:** Deep learning algorithms can extract features of high level and seek to learn from them. At the other hand, the majority of features derived from machine learning are identified by an expert.
- **Time of Execution:** Runtime depends on the various algorithm parameters. Deep learning has greater parameters than algorithms for machine learning. Therefore, the runtime of DL algorithms is more than ML algorithms, especially the training time. However, the DL algorithms 'processing time is less than ML.
- **Problem solving Approach:** Deep learning addresses the end-to-end problem when the machine Learning is used to solve the problem historically, i.e.,, by splitting it into bits.

Basic Terms in Deep Learning

- **Neurons:** Just as a neuron forms our brain's fundamental feature, so a neuron forms the fundamental structure of a neural network. Think about what we do when we obtain new information. We process it and then generate an output when we get the information. Similarly, if a neural network is involved, as shown in the Fig 4.1 the input of a neuron is obtained and processed and generated by an output sent for further processing to other neurons or the final output.
- **Weights:** The input is multiplied by weight as applied to the neuron. For instance, if a neuron has two inputs, the weight is allocated to each input. We randomly initialize weights and change these weights during the model training. After preparation, the neural network gives more weight to the input it finds more important than the less important ones. A weight of zero refers to the insignificance of the particular element.

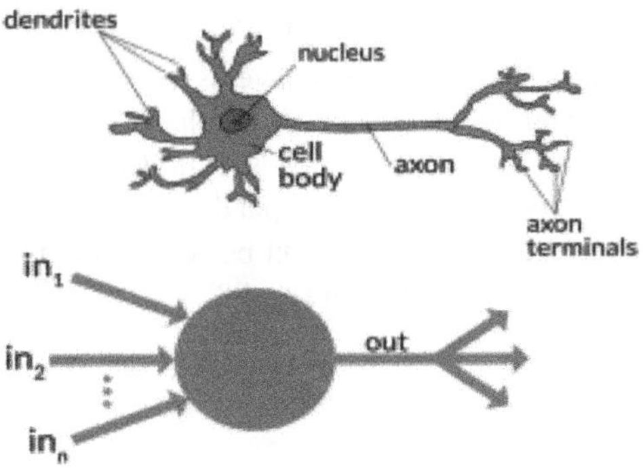

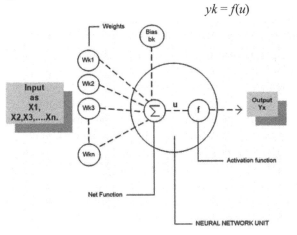

Fig. 4.1 Structure of Neuron

- **Bias:** Apart from the weights, the input is applied with a further linear component, called the bias. The result of weight propagation is added to the input. Basically, the distance of the input multiplied by weight is applied to alter. The outcome will look like a*W1+bias after introducing this bias. This is the final linear dimension of the transformation in the data.

- **Activation Function:** A non-linear function is introduced when the linear component is added to the data. The activation function converts the input signals to output signals. Applies this function to the linear relation. The result after activation will look like $f(a*W_1 + b)$, where f is the activation function.

The inputs given under X_1 to X_n and according to the weight Wk_1 to Wk_n are given as "n" in the diagram below. We've got a preference given as bk. First, weights are multiplied by their respective input and then added together. Let's call this like u.

$$u = \sum w*x + b$$

Accessible for u i.e., $f(u)$ is the activation function and we get the final neural output of

$$yk = f(u)$$

Fig. 4.2 Activation Function

The most commonly used activation functions are sigmoid, ReLU and SoftMax

- **Input/Output/Hidden layer:** As the name suggests, the input layer is the first layer of the network that receives the input. The output layer is the output layer or is the final network layer. The transmission layers of the network is the hidden layers. These secret layers are those that perform different tasks on the input data and transfer the output to the next layer. The layers of entry and output can be seen while the intermediate layers are covered and all this is represented in Fig 4.3.

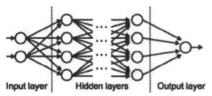

Fig. 4.3 Representation of layers

Input layer | Hidden layers | Output layer

- **Multilayer Perception:** A single neuron cannot accomplish highly complex tasks. Therefore, we generate the necessary outputs with neuron stacks. We have an input, a hidden layer and an output layer on the simplest network. Growing layer has multiple neurons and in each layer are all neurons connected to the next layer of neurons. These networks can also be called fully connected.

- **Cost Function:** The network tries to estimate how close the performance is to the actual value when we create a network. We use the cost/loss function to calculate this precision in the network. If the network makes errors the cost or loss function attempts to penalize.

  Our goal is to enhance our predictability and reduce error while running the network and thus decrease its costs. A minimal cost or loss function is the most efficient performance.

  If the cost function is defined as the medium squared error, it can be written as:

  $$C = 1/m \sum (y - a)^2$$

  where m is the number of training inputs, a is the predicted value and y is the actual value of that particular example. The cycle of learning is about reducing costs.

- **Gradient Descent:** The gradient descent is an algorithm of optimization for cost reduction. To think about it intuitively, you would walk down a hill and go down rather than just run down immediately. So what we're doing in the Fig 4.4, we'll step down from a point x, i.e., delta h, to x- h and we've got to do the same until the edge. To find the local minimum of one function, take steps proportional to the negative of the gradient of the function, consider the bottom as the minimum cost point.

- **Learning Rate:** In each iteration, the learning rate is defined as the level of cost minimization. The rate at which we go down to the minimum of cost function is simply the rate of learning. We should carefully select the rate of learning, as the optimum solution should not be skipped or incredibly low to be converged in the network forever which can be Represented as shown in Fig 4.5.

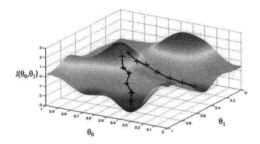

Fig. 4.3 Representation of layers

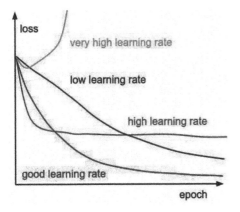

Fig. 4.5 Representation of Learning Rate

- **Back propagation:** We allocate a random weight and bias to our nodes when we construct a Neural Network. We can measure the network error once we have obtained the data for one iteration. This error is then returned to the network with the cost function gradient to correct the network weights. Such weights are then modified in order to reduce the errors in the following iterations. This weight change is called back propagation with the cost function gradient.

  In the meantime, the network movement is reversed, the error and the gage is extracted from the hidden layers from the outer side and the weights are changed.

- **Batches:** During the course of a neural network we divide the input into multiple parts of equal size by chance, instead of sending the entire input in a single go. The batch data training makes the model more scalable when the entire data set is fed in one go into the network as compared with the model developed.

- **Epochs:** An epoch is characterized as a single iteration for all lots of propagation in the forward and backward. That means that one epoch is a reverse pass of the complete input data.

  You can select the number of epochs to train your network. More epochs are more likely to show higher network precision, but it will take longer to converge on the Network. You will need to make sure the network is over-fit if the number of epochs is too small.

- **Dropout:** Dropout is a technique of regularization that avoids network over fitting. As the name indicates, a number of neurons in the hidden layer are lowered randomly during training. This means that training takes place for various neuron combinations in many architectures in the neural network. You may consider dropping out as an ensemble process, which then generates the final result using the output of multiple networks and this is demonstrated in Fig 4.6.

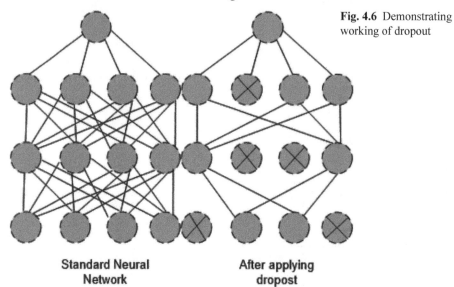

Fig. 4.6 Demonstrating working of dropout

Standard Neural Network   **After applying dropost**

- **Batch Normalization:** Batch normalization can be viewed as a process that we have placed in a river as a special control point. This is done to ensure the data distribution is the same as the next layer. The weights change after each gradient descent phase when we training the neural network. This modifies the way the data form is sent to the next layer. But, the next layer expected the distribution to be like what it had seen before. So, before sending it to the next stage, we specifically normalize the data.
- **Training:** Weights begin as random values, and as the neural network becomes familiar with what input data type is allowed, the network changes weights based on any categorization errors arising from the previous weights. The neural network is referred to as preparation. Once the network has learned, the result for the related input can be predicted by us.

Neural Networks

Neural networks are computer parallel devices which attempt to construct a brain computer model. The main objective is to develop a system which allows several computer tasks to be performed faster than traditional systems. These tasks include patterns recognize and identify, estimate, optimize and cluster data.

Artificial Neural Networks (ANN)

An effective computing system with a core theme taken out of the analogy with biological neural networks are an artificial neural network (ANN). Artificial Neural

Systems, Parallel Distributed Processing Systems and Connectionist Systems are also known as ANNs. In order to allow contact between them, ANN acquires a wide set of units linked in some pattern. Such units are simple processors which work in parallel, also known as nodes or neurons.

Every neuron is linked by a connecting link with another neuron. Each connection link is linked to a weight with input signal information. This is the most useful knowledge for neurons to solve a particular problem, as weight usually activates or inhibits the signal. The inner state of each neuron is called the activation signal. After combining input signals and activation rule, the output signals produced can be transmitted to other units.

Model of Artificial Neural Network:

The diagram below represents the general ANN model and its procedure.

Diagram

For the above general model of artificial neural network, the net input can be calculated as follows:

$$y_{in} = x_1.w_1 + x_2.w_2 + x_3.w_3...x_m.w_m$$

i.e., Net input $y_{in} = \sum_i^m x_i.w_i$

The output can be calculated by applying the activation function over the net input.

$$Y = F(y_{in})$$

Output = function net input calculated

Processing of ANN depends upon the following three building blocks –

Network Topology

The arrangement of a network with its linking nodes and lines is a network topology. ANN can be categorized as the following forms according to topology:

Feed forward Network: It is a non-recurring network with layer processers and all layer nodes are connected to previous layer nodes. The relation has different weights. Feedback loop does not occur so that the signal can only pass from input to output in one direction. The following two forms can be subdivided.

Single Layer Feed forward network: The definition has only one weighted layer of Feed forward ANN. This is, the input layer can be assumed to be fully connected to the output layer.

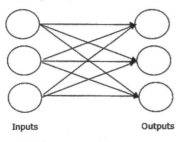

Fig. 4.7 Single Layer Feed forward Network

Inputs   Outputs

Multiple layer feedforward Network: The theory has more than one weighted layer of feedforward ANN. This network is called hidden layers because it has one or more layers between the input and output layer.

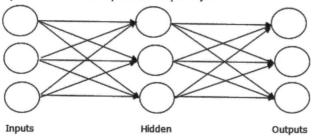

Fig. 4.8 Multiple Layer Feedforward Network

Inputs   Hidden   Outputs

Feedback Network: As the name implies, a feedback network has feedback paths, indicating that the signal will flow through loops in any direction. This makes it a non-linear dynamic system that constantly changes before a balance is reached. The following forms can be separated:

Recurrent Networks: These are closed-loop feedback networks. The two forms of recurrent networks are as follows.

Fully Recurrent Network: This is the most basic neural network architecture because all nodes are connected to all of the other nodes. As can be seen in the below figure Fig 4.9.

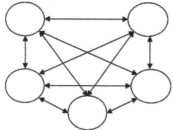

Fig. 4.9 Fully Recurrent Network

Jordan Network: It is a closed-circuit network where the output is transmitted to the input as feedback in the Fig. 4.10 below.

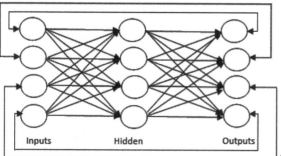

Fig. 4.10 Jordan Network

Inputs   Hidden   Outputs

Adjustments of Weights

or Learning

The method of modification of the weights of connections between the neurons of a given network is learning in an artificial neural network. ANN learning can be categorized into three groups, which include supervised, unsupervised and Reinforcement learning. As we already studied about these learning in before chapter here we will discuss about how ANN uses these learning.

Supervised Learning: This form of education is, as the name implies, under the guidance of a teacher. This is a dependent learning process.

The input vector is introduced to the network during ANN's training under supervised learning, providing an output vector. Compared with the appropriate output vector this output Vector is. If the actual output and the desired output vector vary, an error signal is produced. The weights are modified according to this error signal until the real output matches the desired output as seen in the Fig. 4.11.

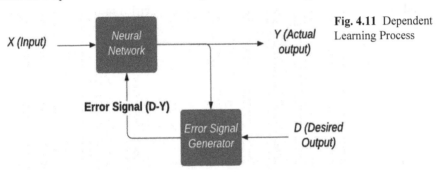

Fig. 4.11 Dependent Learning Process

Unsupervised Learning: This method of learning, as the name implies, is performed without an instructor's supervision. This is an individual learning process.

The input vectors of the same form are coupled with clusters in the training of ANN under uncontrolled instruction. The neural network will provide an output response indicating the class to which an input pattern belongs, if a new input pattern is added.

There is no environmental input as to what the optimal output will be, and whether it is right or wrong. Therefore, as seen in the Fig 4.12 the network itself must discover the patterns and features of input and the relationship between input and output for this form of learning.

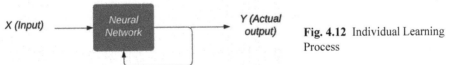

Fig. 4.12 Individual Learning Process

Reinforcement Learning: This kind of learning, as the name implies, is used for strengthening or improving the network through critical data. This learning process is similar to guided learning, but we could have far less knowledge.

The network provides some input from the community during training of the network. This makes the supervised learning very similar. However, the input received here is not descriptive and evaluative, because no instructor exists as in supervised education. The network will change the weights to optimize important information in the future after receiving the feedback as it can be seen in the Fig 4.13.

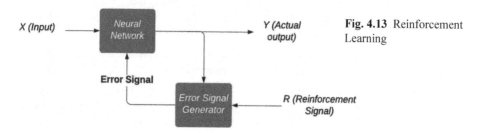

Fig. 4.13 Reinforcement Learning

Activation Functions

The additional force or effort over the input can be specified to obtain an accurate output. In ANN, activation functionality can also be added to achieve the same result through the input. Below are some important activation functions

Linear Activation Function: The identity function is also named as it does nothing to edit the data. It is possible to define:

$$F(x) = x$$

Sigmoid Activation Function: These are of two types namely:

Binary Sigmoidal Function: Input editing takes place between 0 and 1 this activation feature. By fact, it is optimistic. It is always limited, meaning that its performance is not less than 0 and not more than 1. It also increases purely in nature, meaning that the higher the input the output. You may describe it as

$$F(x) = sigm(x) = 1 \div 1 + exp(-x)$$

Bipolar Sigmoidal Function: This activation function adjusts the input from -1 to 1. In fact, it can be positive or negative. It is always limited, meaning that its production cannot be below -1 and above 1. This also rises purely in nature such as sigmoid. It can be set as

$$F(x) = sigm(x) = 2 \div 1 + exp(-x) - 1 = 1 - exp(x) \div 1 + exp(x)$$

Challenges in ANN

The first step in resolving the question of an image classification using ANN is to transform a 2D image into a 1D vector before training. There are two inconveniences:

- When the image size increases, the number of trainable parameters increases dramatically
- ANN loses the image's spatial characteristics. Spatial characteristics apply to the pixel arrangement in a image.

ANN cannot record sequential information for the processing of sequence data in input data.

Convolutional Neural Network

Convolution neural network is the same as regular neural network as it consists of neurons with learning weights and prejudicial characteristics. Normal neural networks ignore the input data structure and convert all data to a 1-D array before they are transferred into a network. This method corresponds to standard data, but the process can be tedious if the data includes images. This problem is easily solved by CNN. It takes into account the 2D structure of the images when processing them, enabling them to extract image-specific properties. The main aim of CNNs is therefore to transfer from raw image data in the input layer into the correct class in the output layer. The only difference in the treatment of incoming data and layer sort is between ordinary NNs and CNNs.

Architecture Overview of CNNs

The standard neural networks are architecturally assisted with an input and transform it through a hidden layer. With the support of neurons, each layer is related to the other layer. The main drawback of ordinary neural networks is that they do not reach full images correctly.

The CNN architecture is structured with neurons called width, height and depth in three dimensions. The current layer of each neuron is linked to a small piece of the preceding layer output. The overlay of a filter N to N on the input image is identical. M filters are used to ensure that all the information is obtained. These M filters are extracting elements such as sides, corners, etc.

Layers Used to Construct CNN

- **Input Layer:** It takes the raw image data as it is.
- **Convolutional Layer:** This layer is the central component of the CNNs where most computations are performed. This layer measures the convolutions in the input between the neurons and the specific patches.
- **Padding:** Filter also doesn't suit the input image perfectly. We have two alternatives:
  - Pad photos with zero padding (zero padding) to fit.
  - Delete the picture component that didn't match the filter. This is known as a valid padding that only holds valid part of the image
- **Rectified Linear Unit Layer:** It is used to activate the previous layer output. It increases the network's nonlinearity to allow for any form of function to be well generated. Because real-world data are non-negative linear values, our Network will want to know.
- Other non-linear functions are also possible, for example tanh or sigmoid, instead of ReLU. Many data scientists use ReLU because the ReLU is better than the two others.
- Pooling Layer: Pooling just helps to preserve the essential parts of the network as we progress. The pooling layer acts on each input slice independently and spatially resizes it. Spatial pooling, which reduces the dimensionality of each map but retains essential details, is often called subsampling or down sampling. Different forms of spatial pooling may be:

- Max Pooling
- Average Pooling
- Sum Pooling

Max pooling takes from the corrected feature map the largest element. The average pooling will take the biggest element as well. Description of all elements in the function map

- Fully Connected Layer/Output Layer: We flatter our matrix and feed it to a completely connected layer such as a neural network, as a layer called an FC layer. The output values in the last level are calculated by this layer. The resulting output is $1 \times 1 \times 1 \times L$ in which L is the dataset number training groups.

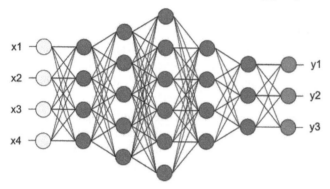

Fig. 4.14 After pooling layer, flattened as FC layer

The function map matrix is transformed as a vector in the above Fig 4.14 (x_1, x_2, x_3, etc). We combined these features to create a layout with the fully connected layers. Eventually, we have an activation function for classifying outputs, such as SoftMax or sigmoid.

We can clearly understand the full architecture of CNN from the Fig 4.15.

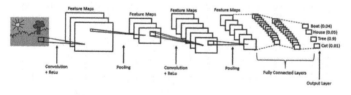

Fig. 4.15 Complete CNN architecture

Recurrent Neural Network

A recurring neural network (RNN) is an artificial neural network class, where a direct graph along a sequence relation forms a unit. This allows complex time actions to be exhibited for a period of time. As compared to neural feed forward networks, internal memory of RNNs may be used to process input sequences. This makes them applicable to tasks like the unsegmented, related recognition of handwriting or language.

If every model requires a context to provide the output based on the data, the recurrent Neural Network comes into action. The context is often the most important thing for the model to predict the best performance.

Let's get an analogy to understand. Suppose you watch a movie, you continue to watch it because at all times, you have a history as you watched the movie before so, then you can only relate it all correctly. This means you know what you've been watching.

Similarly, RNN remembers everything. Both inputs and output are distinct from each other in other neural networks. In RNN, however, all inputs are connected. for instance, the relationship between all the terms helps to predict the next word in a given sentence which can lead to better output. During the training itself, the RNN recalls all these ties.

To do this, the RNN creates networks with loops that allow information to persist as in Fig 4.16.

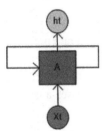

Fig. 4.16 RNN with a LOOP

The neural network will sequence input from this loop structure. You'll understand it better if you see the unrolled structure of network in the Fig 4.17.

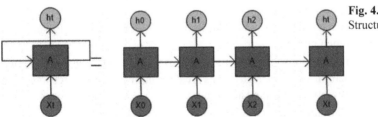

Fig. 4.17 Unrolled Structure of RNN

As in the unrolling structure you can see. First of all, the $x(0)$ is taken from the input sequence and then $h(0)$ is generated, which is the input together with $x(1)$. The input for the next step is h(0) and $x(1)$. Similarly, $h(1)$ for the next step and so on is the entry with $x(2)$. It preserves its meaning during training.

"Whenever there is a data series and the temporal dynamics connecting the data are more important than the spatial content of each frame."

Types of RNN's

The key explanation for the more thrilling recurring networks is that it allows one to operate on sequences of vectors: input sequences, output sequences or in most cases both. This could be more specific with a few examples:

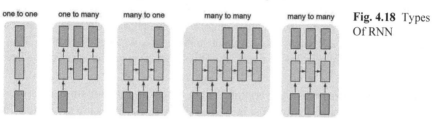

Fig. 4.18 Types Of RNN

Each rectangle in above image represent Vectors and Arrows represent functions. Input vectors are Red in color, output vectors are blue and green holds RNN's state.

One to One: It often named Plain neural networks. It addresses Fixed input size to Fixed output size where it is independent of previous data/output.

Example: Classification of images.

One to many: This deals with information of a fixed size as input and provides data sequence as output.

Example: Image tagging takes the image as input and shows a word sentence.

Many to One: It takes information sequence as an input and passes a certain output scale.

Example: sentimental analysis where the sentences is positive or negative is graded.

Many to Many: It takes an information sequence as input and processes it repeatedly produces a data set.

Example: Machine translation where an RNN reads an English phrase and then publishes a French phrase.

Bidirectional Many to many: Input and output synced sequence. Note that no pre-specified length sequence restrictions are present in any case, since repeating (green) transformations are fixed and can be applied as much as we want.

Example: video grading where each video frame is named.

Challenge of RNN

Deep RNNs (RNNs with several time steps) are often affected by the disappearance and exploding problems in gradients which are typical in all the neural network types.

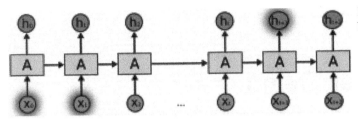

Fig. 4.19 Recurrent Neural Network

As you can see in the Fig 4.19, the calculated gradient will disappear when the initial time stage is reached.

Difference Between ANN, CNN & RNN

We will depicts all the differences between Artificial Neural Network, Convolution Neural Network and Recurrent Neural network using a table for better understanding.

	ANN	CNN	RNN
Data	Tabular data	Image data	Sequence data
Recurrent Connections	No	No	Yes
Parameter Sharing	No	Yes	Yes
Spatial Relationship	No	Yes	No
Vanishing and Exploding Gradient	Yes	Yes	Yes

Summary

In this chapter we have studied about the basics of deep learning which includes the terminology or the basics terms we address while using the deep learning or any other algorithms like bias, learning rate etc., and we have also studied about the artificial neural network, Convolutional Neural Network and Recurrent Neural Network with their basics structure which can be used in different scenarios according to the requirement of the problems. In the Implementation part of this book we will be using CNN for classification and try to recognize the plant or their species.

Chapter 5

Computer Vision

Introduction

Agriculture has played an important role in the global economy in recent years. The further growth of the population is leading to a gradual decrease in the area of the agricultural land and further growing stresses on the farm system. Agricultural food production methods are rising in demand for successful and healthy. Innovative sensoring and motive technologies and advanced information and communications systems and Artificial Intelligence must supplement conventional agricultural management methods to speed up farm productivity growth more accurately, thereby promoting the production of high-quality and high-performance agriculture. Computer vision monitoring systems have become valuable tools in farm operations over the past decades, and their use has increased significantly.

What is Computer Vision?

Computer vision can be characterized as a field of AI for extracting information from digital images. The type of data obtained from an image may be different from recognition, navigational space measurements or applications of increased reality.

The applications of computer vision can also be described. Computer vision builds algorithms that recognize and use the quality of pictures for other purposes.

People look at and physically experience the world around them through the eyes and the brains. Computer vision is the science that wants to provide a machine or a device with similar, if not better, performance.

The aim of computer vision is to automatically collect, interpret and understand useful and meaningful data from a single image or a series of images. This consists of establishing a theoretical and algorithmic basis for automatic visual comprehension.

In agriculture, Computer vision systems are clearly feasible for grading and evaluating properties such as color, form, thickness, surface defects and contamination of food products into different grades.

Computer Vision hierarchy: Computer vision is classified into three basic categories:

- Low Level : It contains a process image for extraction of features
- Intermediate Level: This involves recognition of objects and 3D scenes
- High level: It provides a conceptual overview of an incident, goal and comportment scenario

Task of Computer Vision

- **Recognition:** The normal issue in computer vision, image processing and machine vision is whether or not the image data contains any specific object, function or operation. Normally, a person can overcome this dilemma effectively and easily, however in the general case machine view it is still not satisfactorily solved: arbitrary objects in arbitrary circumstances. The available methods of dealing in this field can at best solve this problem only for particular objects, such as basic geometric objects (e.g., polyhedrons), human features, written or manuscript characters or vehicles, and surroundings in unique circumstances that are usually well described in terms of illumination.

There are different variety of Recognition come to notice like

Recognition: One or more objects or classes, usually together with their 2D image location or 3D poses, can be recognized pre-specified or learned in a given scene. It can be any person, item like fruits, vegetables etc.

Identification: A separate instance is recognized for an object. *Examples*: identification of the face or fingerprint of a specific individual or identification of a particular vehicle.

Detection: For a specific situation the image data is screened. *Examples*: identification, in agricultural images, of potential irregular cells or tissues or automatic road toll identification. Detection by comparatively simple and quick calculation is often used to identify small regions with interesting data, which can be analyzed further in order to obtain a correct interpretation by using more computationally challenging technologies.

- **Motion Analysis:** Various tasks involve the movement estimation when processing an image sequence to measure the speed at each point either in the image or the 3D image scene, or even the camera that produces the images.

Examples are the following:

Egomotion: Determination of the 3D rigid motion of the camera from the camera image series (rotation and translation).

Tracking: Following (normally) movement of a series of points of interest or objects in the picture sequence (e.g., cars, human beings or other organisms).

Optical flaw: to evaluate how the point moves in relation to the image plane for each point in the image, i.e., its apparent motion. This move represents both the way the 3D point moves in the scene and the way the camera moves in relation to the object.

- **Scene Reconstruction:** If a scene reconstruction has one or (typically) more images of a scene or a picture, it is intended to compute a 3D model of the scene. The model may be a series of 3D points in the simplest case. More detailed methods produce a complete 3D surface model. The introduction of 3D imagery without motion or scanning and the related processing algorithms allows fast progress in this area. 3D grid sensing can be used in multi-angle processing of three-dimensional images. Several 3D images are now available to be merged into point clouds and 3D models

- **Image Restoration**: The goal of image restore is to eliminate noise from photographs (sensor noise, motion blur, etc). Diverse filter types such as low-pass filters or medium filters are the easiest possible solution for noise reduction. More advanced approaches are designed to differentiate between noise and local image structures. When the image data are first analyzed by local image structures, such as lines or edges, and then the filtering is managed based on local information from the analysis stage, a better noise removal level is usually achieved compared with simpler approaches.

The recognition/detection of images is correlated with many challenges:

- viewpoint variation
- scale variation
- intra-class variation
- image deformation
- image occlusion
- illumination
- background clutter

Computer Vision System: Computer vision system structure is highly important. In some cases the device is a subsystem of a larger design, for example with subsystems for mechanical actuator control, planning, repositories of information, man-machine interfaces, etc. Some systems have separate implementations, which solve a particular measuring and detection issue. This would also depend on whether its functionality is pre-specified or whether part of the computer vision program can be updated or amended during service. However, many computer vision systems have traditional features.

Image Acquisition: Capturing of One or more images using digital image sensors which are developed to include range sensors, tomography equipment, radar, ultrasonic cameras, etc., in addition to specific type of light sensitive cameras. The resulting image data depends on the form of sensor: a specific 2D image, a 3D volume or a series of images. The pixel values usually refer to the intensity of light in one or more spectral lengths (gray images or color images), but can also apply to various physical measurements, such as sonic or electromagnetic wave width, absorption or reflectance, or the nuclear magnet resonance.

Pre-Processing: In order to retrieve certain pieces of information, a Computer vision approach may usually be used to analyze the data to ensure that all expectations suggested by the technique can be fulfilled.

Feature Extraction: The image data extracts feature at various levels of complexity. Typical examples are the following: lines, edges and ridges. Localized points of interest like edges, blobs or lines. The texture, form or movement can be more complex.

Detection/Segmentation: There is a decision at some stage in the course of the processing which image points or image areas are important for further care. which can be like:

- Order of a particular set of points of interest.
- Segmentation of one or more parts of the image containing a specific point of interest.
- Splitting the image into the nested scene Architecture consisting of the first-ground, object classes, individual objects or salient object pieces (also referred to as the hierarchy of the spatial taxon scene).
- segmenting or co-segmenting one or more videos to a set of foreground masks per frame while preserving its time semantical continuity

High Level Processing: The input typically consists of a small data set, such as a set of points or a picture area which is supposed to contain one particular object. It deals with

- Verification that the data meet clear assumptions based on model and implementation.
- Application-specific parameters such as object position or object size are estimated.
- Image recognition – grouping into various categories of a detected object.
- Image registration – comparing and integrating the same object's two different views.
- **Decision Making:** Giving the final judgment for the submission,
  - *Example*, Automated testing applications pass/fail.
  - Match/no-match in applications for recognition.
  - Additional human review flag in applications for medical, military, protection and recognition.

For the Implementation of Computer vision, we use framework like OpenCV. Which we will study in next section.

OPENCV

OpenCV is an opensource package or library which is developed by Intel. This library is cross platform it means we can use it with java, C++, python etc., In all kind of image and video processing and analysis, the OpenCV or the Open Source Control Vision software (http:/opencv.org) is used. It can process images and videos to recognize people, faces, or even manuscripts. Integrated into many libraries, including NumPy, Python is able to analyze the OpenCV array structure. OpenCV converts visual knowledge machine vision to the space of the object. The image pattern and functionality from the vector space can be defined and mathematical operations are carried out on these features.

Installing OpenCV

On Windows: Below is a list of the steps needed to complete the installation.

Install OpenCV: Run the command "pip install opencv_python" On your Command prompt of windows and OpenCV will be automatically get installed.

On linux/unbuntu:

Importing OpenCV: Get to the IDLE and import OpenCV:

```
>>> import cv2
You can also check which version you have:
>>> cv2.__version__
'3.4.1'
```

Note: Version can be different it depends on its release

Let's try something with OpenCV:

Example save it as opencv.py:

```
import cv2 #OPENCV library
# Load an color image
img = cv2.imread ('download.jpg')
#Image should be in the same folder where program file is saved
cv2.imshow ('image',img)
cv2.waitKey (0)
cv2.destroyAllWindows ()
```

Output:

Fig. 5.1. Output Image

We can convert the above Fig 5.1 image into grey scale by passing value 0 into the argument img

```
import cv2 #OPENCV library
# Load an color image
img = cv2.imread ('download.jpg',0)
# passing 0 to convert it into grey scale
#Image should be in the same folder where program file is saved
cv2.imshow ('image',img)
cv2.waitKey (0)
cv2.destroyAllWindows ()
```

Output:

Fig. 5.2 Grey Image

Explanation: Cv2.imread () is used to read an image. And as seen by passing second argument we have converted image to grey scale as seen in Fig 5.2.

The second arguments always depict how the image should be read:

- **cv2.IMREAD_COLOR:** Loads an image color. Any image transparency is overlooked. It's the predetermined mark.
- **cv2.IMREAD_GRAYSCALE:** Loads image in grayscale
- **cv2.IMREAD_UNCHANGED:** Loads image as such including alpha channel.
- **cv2.imshow ():** used to display an image. The output window automatically fit to the size of an image. The first argument is a string window name. Our photo is the second statement. You can build as many windows, but with different names of windows as you want.
- **cv2.waitKey ():** The keyboard binding feature in OpenCV. The time in milliseconds is the point. For each keyboard case, this feature is waiting for milliseconds. The software continues if you click a key during that time. If 0 is reached, it waits for a central stroke forever. You may also set it to detect other keystrokes such as, when key an is pressed, etc.
- **cv2.destroyAllWindows ():** It destroys all the windows created and if any particular window is to be destroyed, we can use the function cv2.destroywindow () and pass the window name as an argument into it.

OpenCV is widely used in so many domains as an important part of various technologies and we will use open cv in our further chapters for various purpose and a lot of other topics will be discussed. Now let's make an real time project with the help of OpenCV.

Crop Grading

In agriculture crop quality grading is a very important thing as it provides the simple by using OpenCV

```
import cv2
import numpy as np
from matplotlib import pyplot as plt

def get_ classificaton (ratio):
        ratio =round (ratio,1)
        toret=""
        if (ratio>=3):
                toret="Slender"
        elif (ratio>=2.1 and ratio<3):
                toret="Medium"
        elif (ratio>=1.1 and ratio<2.1):
                toret="Bold"
        elif (ratio<=1):
                toret="Round"
        toret=" ("+toret+")"
        return toret
#rnjn
print ("Starting")
img = cv2.imread ('rice.png',0) #Path where images is stored
#load in greyscale mode

#convert into binary
ret,binary = cv2.threshold (img,160,255,cv2.THRESH _ BINARY)# 160 -
threshold, 255 - value to assign, THRESH _ BINARY _ INV - Inverse
binary

#averaging filter
kernel = np.ones ( (5,5),np.float32)/9
dst = cv2.filter2D (binary,-1,kernel)# -1 : depth of the destination
image

kernel2 = cv2.getStructuringElement (cv2.MORPH _ ELLIPSE, (3,3))

#erosion
erosion = cv2.erode (dst,kernel2,iterations = 1)

#dilation
dilation = cv2.dilate (erosion,kernel2,iterations = 1)

#edge detection
```

```
edges = cv2.Canny (dilation,100,200)

### Size detection
_,contours,hierarchy = cv2.findContours (erosion, cv2.RETR_
EXTERNAL, cv2.CHAIN_APPROX_SIMPLE)
print ("No. of rice grains=",len (contours))
total_ar=0
for cnt in contours:
            x,y,w,h = cv2.boundingRect (cnt)
            aspect_ratio = float (w)/h
            if (aspect_ratio<1):
                        aspect_ratio=1/aspect_ratio
            #print (round (aspect_ratio,2),get_classificaton
                                          (aspect_ratio))
            total_ar+=aspect_ratio
avg_ar=total_ar/len (contours)
print ("Average Aspect Ratio=",round (avg_ar,2),get_classificaton
                                          (avg_ar))
#plot the images
imgs_row=2
imgs_col=3
plt.subplot (imgs_row,imgs_col,1),plt.imshow (img,'gray')
plt.title ("Original image")

plt.subplot (imgs_row,imgs_col,2),plt.imshow (binary,'gray')
plt.title ("Binary image")

plt.subplot (imgs_row,imgs_col,3),plt.imshow (dst,'gray')
plt.title ("Filtered image")

plt.subplot (imgs_row,imgs_col,4),plt.imshow (erosion,'gray')
plt.title ("Eroded image")

plt.subplot (imgs_row,imgs_col,5),plt.imshow (dilation,'gray')
plt.title ("Dialated image")

plt.subplot (imgs_row,imgs_col,6),plt.imshow (edges,'gray')
plt.title ("Edge detect")
```

```
plt.show ()
```

Output:
```
>>>
RESTART:        E:\studies\Projects\BOOK\Rice\rice-quality-analysis-
master\code.py
Starting
No. of rice grains= 30
Average Aspect Ratio= 2.3 (Medium)
```

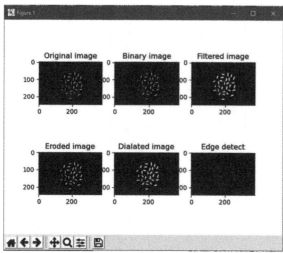

Fig. 5.3 Output of Rice using Matplotlib

As we can see in the Fig 5.3 how rice has been detected and grading has been done using OpenCV which is represented using matplotlib library which is used for visualization of output after grading of rice.

Summary

Computer Vision is considerably the eyes of artificial Intelligence which can help to process any video or pictures and understand them for generating a rightful data. In this chapter we have seen the importance of computer vision and how it can be useful in the agriculture for identification and with two small example we have also demonstrated the use of computer vision.

Chapter 6

Knowledge Based Expert System

In this chapter we will look at the Expert system which can guide and assist an individual once it is provided with the enough Knowledge to do so. Here we will discuss about expert system and its architecture with proper steps of making an expert system in any field by several tools which are available to do so. For expert systems the most important is Knowledge Engineering which is the collection of Knowledge from an expert and Engineering it in such a form that a machine could understand it.

Introduction

One of the pioneering fields of AI study is Expert Systems (ES). It is proposed by researchers in the computer science department of Stanford University. An expert system is a computer system which emulates a human expert's decision-making skills. Expert systems are developed by reasoning through information structures, representing laws, rather than through traditional procedural code, in order to overcome complex problems.

Expert Programs use professional expertise systematically to address human expert problems. An expert is an individual with experience in a specific field. The expert has expertise or special credentials which most people do not learn or care about. An expert can solve or solve problems more effectively than people can solve. Expert System actually emulates all these qualities of expert and solve the problem in a better way and fast.

Today, Expert System (ES) is used in the fields of finance, research, engineering, technology, medicine and many others where there is a well-defined problem area. The fundamental principle is that an expert method should also define the steps to illustrate why a problem can be solved. Unless an individual can justify his reasoning for Expert System.

Expert System always works on basis of its problem. In expert System designing we always define the problem domain and latter on we will collect knowledge related to it. As you can see in the below figure knowledge domain of an Expert system is always a subset of problem domain as represented in the Fig 6.1. The knowledge which is being built is always gathered from books, magazines, papers or experts based on their experience in particular field.

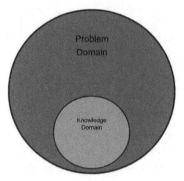

Fig. 6.1 Expert System Knowledge domain

All the knowledge which is gathered will be based on the problem defined and it will specific and based on facts.

Characteristics of Expert System

Knowledge is the most critical aspect of any expert program. The power of the expert system lies in the high-quality awareness of task areas. In expert systems, knowledge is separated from its processing i.e., the knowledge base and the inference engine are split up. For storing this information, a standard system is a mixture of information and the control structure. The combination creates issues with the system code's comprehension and analysis, as any alteration in the code affects information and processing. The expert method includes a knowledge base and a set of rules to apply the knowledge base to each unique situation defined in the program. Comprehensive expert systems may be improved by adding to the knowledge base or the collection of rules. Expert system can be developed from scratch or designed with a software development part called a "kit" or shell. A shell is a full development environment for the creation and maintenance of application based on knowledge. It offers a step-by-step approach and an preferably user-friendly interface for a knowledge engineer who can directly involve domain experts in structuring and coding the information

Expert system as seen can be developed and they have some general characteristics which are

- **High Performance:** The system must be capable of responding at very high level of competency or better than that of any expert in the field. i.e.,, the advice quality given by the system should be high and better as compart to the expert.

- Adequate Response time: The Expert should perform or response to a situation quicker as compared to expert and with an accurate and better advice or solution to the query.

- **Good Reliability:** The expert system should not be prone to lags and crashes it should be well functioning in any situation.

- **Understandable:** The Expert system should able to explain its way of decision making on query so that it can be understandable.

- **Flexibility:** The Expert System contains large amount of data in the form of knowledge regarding or related to certain field and it should be very easy to add, change and delete any knowledge into the system for its improvement.

Architecture of an Expert System

An expert system tool, or shell, includes the essential components of expert systems. The knowledge base and the Inference engine are the main components of the expert systems. and other Elements of Expert System which can be seen in the Fig 6.2 are:

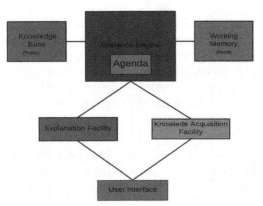

Fig. 6.2: Architecture of Expert System

- **Knowledge Base:** The knowledge base includes the facts to be learned, formulated and solved. It is a domain information store captured by the human expert through the module for learning. It uses rules, frames, logic, semantic network, etc to reflect the information output. The expert system knowledge base provides empirical and heuristic knowledge. True knowledge is the general knowledge of the task domain, commonly contained in textbooks or journals. Heuristic information is a less rigorous, experiential, judgmental, and generally individualistic awareness of success. It includes knowledge of good practice, good judgment, and logic possible in the area.
- **Inference Engine:** It is an expert System's brain. It uses the control system and offers a reasoning technique. This serves as an interpreter who analyzes and operates on the rules. This is used to suit histories of the responses and firing rules received by users. The key task of the inference engine is to map the path to a conclusion through a set of rules. Two methods, i.e., forward and rear chaining, are used here.
- Working Memory
- Agenda: It contains the list of the task or queries which need to be performed by expert system.
- **Explanation facility:** It is a subsystem which explains the behavior of the system. The description can differ from how to arrive at the final or intermediate solutions to explain the need for further detail. Here users want to ask simple questions as to why and how the device information is being shared with the user and acts as a tutor.
- **Knowledge Acquisition Facility:** The acquirement of knowledge is a buildup, transition and transformation into a computer system to create or expand the

knowledge foundations of problem-solving expertise from experts and/or recorded information sources. It is an expert subsystem for developing information bases. For the acquisition of information, process analysis, interviews and observation techniques are used

- **User Interface:** It is an interaction device with the user. It provides facilities for user-friendly conversation, such as menus, graphical gui etc. The User Interface is responsible for translating rules from its internal (which users do not understand) representation into user comprehensible types.

Knowledge engineering is known to develop the expert System. Domain experts, users, knowledge engineers and system maintenance staff are employees involved in the expert system development. Field specialists have unique expertise, assessment, practice and strategies for offering guidance and solving problems. It provides information on job success. During the creation of the inference engine, the knowledge base framework and the user interface, the knowledge engineer is involved. The Knowledge Engineer and the specialist in information will foresee the needs of the consumer in creating a program

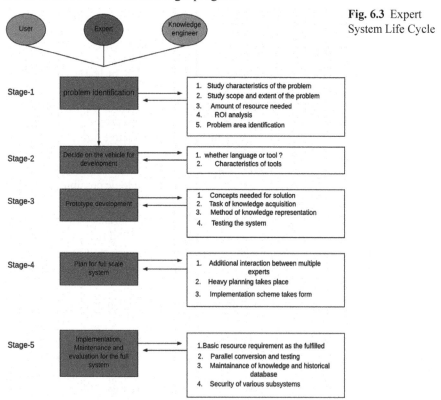

Fig. 6.3 Expert System Life Cycle

The development of an expert system includes five key phases as seen in the Fig 6.3. Each stage has its own unique characteristics and is associated with other stages.

Stage 1: Problem Identification: In this step, the expert and the Knowledge Engineer communicate to detect the problem. The key issues previously discussed are investigated for the problem's characteristics. The degree and nature are examined. The quantity of resources available, such as people, computer resources, finances etc. The ROI analysis is conducted. Problem areas that can cause a great deal of trouble are defined and the overall design and conceptual solution to that problem is established.

Stage 2: Deciding Mode of Development: The immediate phase, when the problem is found, is to determine which model will be built. The software technician may use a programming language like PROLOG and LISP or some modern language to develop the framework or use a development shell from scratch. During this point, different shells and instruments for their suitability are described and analyzed. Tools whose properties suit the problem are thoroughly evaluated.

Stage 3: Prototype Development: The following are the required activities before creating a prototype-
- Decide on which concepts the solution would have to produce. The level of information (granularity) is an essential factor to be calculated. The development of the system starts with coarse granularity and progresses to fine granularity.
- Then starts the process of gaining information. The knowledge engineer and the field expert often communicate, and domain information is extracted.
- The knowledge engineer determines the form for representation until the expertise has been acquired. A conceptual image of the representation of information should have emerged during the identification process. This view is implemented or updated in this process.
- A prototype is developed when the scheme of information representation and information is available. Each prototype is evaluated for different problems and the prototype is reviewed. This method results in knowledge of fine granularity, which is effectively encrypted in the basis of information.

Stage 4: Full System planning: The success of the prototype gives the full-scale device impetus. The region of the problem that can be fairly easily applied is first selected in prototype design. The creation of a subsystem is allocated to group leaders during full implementation, and schedules are drawn up. Gantt map, PERT or CPM are to be used.

Stage 5: Implementation, Evaluation and Maintenance: It is an expert system's final life cycle process. At the site, the entire scale system is created. The basic criteria for services are met at the site and simultaneous conversion and testing strategies are adopted. The final device is rigorously checked and ultimately passed to the customer.

Evaluation in any AI system is a challenging task. Solutions to AI problems are, as already described, only satisfactory. Since the evaluation requirements are not available, evaluation is difficult. What can be done is to provide the program and human expert with a series of problems and to compare the results.

System maintenance requires changing the knowledge base since there is no established information, environment, or issues. It is important to maintain the historical records and keep track of minor modifications made to the Inference engine. Maintenance also includes security.

Tools for Development of Expert system

There are some tools which are available for development of an expert system as represented by fig 6.4. These tools are divided into four categories which are:

(a) Programming Languages

(b) Knowledge Engineering Languages

(c) System Building Aid

(d) Support facilities

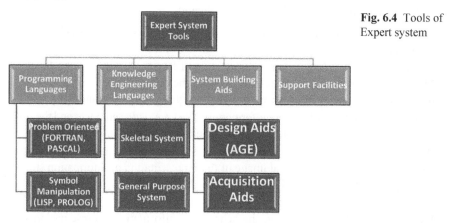

Fig. 6.4 Tools of Expert system

(a) **Programming Languages:** There are certain languages which are available and can be used to develop an expert system. It always based on the type of the problem we have like

   i. **Problem-Oriented languages:** Such as Fortran and pascal. These are designed for a particular class of problems. Fortran have a convenient feature for performing algebraic calculations.

   ii. **Symbolic Manipulation Languages**: such as LISP and PROLOG. LISP has the mechanism which can manipulates the symbol available in the form of list structures.

(b) **Knowledge Engineering Languages:** A language of knowledge engineering is a sophisticated method to build systems for experts consisting of an expert system that builds language in an extensive support environment.

   i. **Skeleton System:** A skeletal language of information creation is a simplified expert system. Domains are removed from the expert program, leaving only the Inference engine and support facilities. For example, MYCIN is tripped down to EMYCIN which is easy and fast.

ii. General Purpose System: A Knowledge Engineering language can handle many problem areas and forms for a general purpose. This offers more power than skeleton framework over data access and search, but is much harder to use.

It varies according to the flexibility and generality. The Skeletal and General-purpose system falls in the research system category.

(c) **System Building Aids:** Consists of programs that help to develop the expertise of the industry expert and programs that help design the Expert System construction method.

i. Designing Aid: It helps in designing the expert system.

For example: Like age

It Assist the design and development of an expert System by the knowledge engineer.

HEARSAY, first used in the mid-1970s, interprets the enemy data on radio contact HEARSAY-III HANNIBAL. This uses data position information and signal characteristics to classify organizational units and their order of battle.

Provides the consumer with a collection of components that can be assembled as building blocks to form parts of an expert structure.

An expert system structure like forward chains, backward chaining, or blackboards architecture supports each component, a compilation of INTERLISP functions.

ii. Knowledge Acquisition Aid: It helps in Knowledge acquisition for the Expert system.

For example: TEIRSIAS

Its system design helps transfer information to a knowledge base from a domain expert.

Only used for management of the database.

The program learns new rules for the problem area through an interaction allowing users to define rules in a specific subset of English.

This analyzes the rules, determines how accurate and clear these rules are and lets the user fix them.

(d) **Support Facilities:** Consist of tools for helping programming – like debugging aids, Knowledge base editors to enhance capabilities of a finished system and Extra software packages.

The main components of Support Facilities are

i. **Debugging aids:** It is divided into different types like

Trace facilities: shows name or rule no of fired rules list, sub-routines called.

Break packages: allow the user to tell where to stop program execution so that the user can stop it just before an error occurs.

Automated testing: allows the user to test a program automatically on several benchmarks

i. I/O facilities: It is divided into different types like

Run time Knowledge Acquisition: the tool itself enables the user to converse with the system and to seek the information required in the knowledge base.

Menus: provide the user with the menus to choose for inputs. Menus: No specific code required.

Accessibility of the operating system: enables ES to track and manage other work during service

(e) **Explanation facilities:** It is divided into different types like

Retrospective: it explains how the system reached a particular state. How system arrived at a conclusion

Hypothetical: explain what would have happened if certain fact or rule had been different.

Counterfactual: why an expected conclusion has not reached

(d) **Knowledge base editors:** It is divided into different types like

Automatic book keeping: It helps to monitor the changes made by the user.

Syntax checking: It checks whether the rules are entered in correct format and are in accordance with the grammatical structure of ES languages.

Consistency checking: It checks the semantics of rule entered to check if there is any conflict with the existing knowledge base.

Knowledge Extraction: This helps to add new rules or modify the rules

Pros and Cons of Expert System (ES)

It has a lot of attractive feature.

- **Increased Availability:** It increases the availabilities of its expertise in mass on any suitable computer hardware.
- **Reduced Cost:** The cost of expertise providing every user is decreased a lot.
- Reduced Danger: It can be utilized in any hazardous or remote location.
- **Permanence:** Once the expert system is developed it is permanent whether expert of that domain can be lost but expert system will last ever.
- **Multiple Expertise:** The Knowledge of several Expert system can be made work simultaneously and regularly on a given problem until it gets solved at any time of night or day. The level of expertise of an expert system is much higher than the human expert of that domain.
- **Increased Reliability:** Expert system are steady, unemotional and always complete the response at time.
- **Explanation:** The Expert System provide all the description how it reached to the conclusion on a query or problem. In the same case human expert cannot provide all the time explanation due to its tiredness or ill conditions.

- **Fast Response:** Expert system is capable fast and real-time responses all the time.
- **Intelligent tutor:** Expert system can become a tutor for student to study apply basic programs and explains its outcome.
- **Intelligent database:** It can be used to access databases intelligently.

Cons of Expert System

- Expert system has shallow Knowledge.
- It has no deep understanding of the concepts and their relationship until we provide.
- It doesn't have common-sense
- It is a closed world with specific knowledge.
- Expert system may not have or select the most appropriate method for solving a particular problem.
- Some Easy problem in expert system are computationally becomes expensive

Summary

Expert System has become a very pioneering branch of Artificial intelligence with several uses in several fields. After studying about expert system, we can see what is an expert system, how expert systems are built and what are the tools for making an expert system. Expert system in agriculture field can help the farmers in lot many ways by guiding them with proper knowledge to use advance technology and other herbal to increase their crop production and protect it from pests.

PART B: IMPLEMENTATION OF ARTIFICIAL INTELLIGENCE

Chapter 7

Tools for Artificial Intelligence

In this chapter we will talk about the tools which are required to make project in artificial intelligence. All this bellow discussed tools are uses python programming language for making any project in any respective field. We have discussed about how anyone can install and use the tools to built a real time project in artificial Intelligence.

Using Python IDE

To download and install Python visit the official website of Python https://www.python.org/downloads/ and choose your version. We have chosen Python version as per requirement all the codes in this book were on python version 3.6 as you can see in the Fig. 7.1.

Fig. 7.1 Python Download

Once the download is complete, run the exe for install Python. Now click on Install Now as in Fig. 7.2.

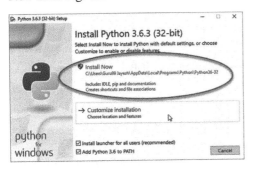

Fig. 7.2 Installation of Python

You can see Python installing at this point as in Fig. 7.3

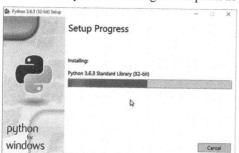

Fig. 7.3 Installation process

When it finishes, you can see a screen that says the Setup was successful. Now click on "Close" as in Fig. 7.4.

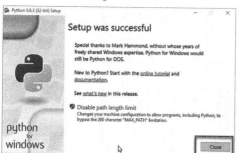

Fig. 7.4 Installation Complete

Using Anaconda

On windows:

Go to the following link: Anaconda.com/downloads.

The Anaconda Downloads Page will look something like this in Fig. 7.5:

Fig. 7.5 Installation Complete

Select which ever your operating system is from the three listed operating systems as in Fig. 7.6. we have chosen windows.

Fig. 7.6 Anaconda Versions Select windows

Download the most recent Python 3 release as in Fig. 7.7. At the time of writing, the most recent release which I was using is Python 3.6 Version. Python 2.7 is legacy

Python. For problem solvers, select the Python 3.6 version. If you are unsure if your computer is running a 64-bit or 32-bit version of Windows, select 64-bit as 64-bit Windows is most common.

Fig. 7.7 Python version for Anaconda

The download is quite large so it may take a while to for Anaconda to download. Once the download completes, open and run the.exe installer. At the beginning of the install, you need to click Next to confirm the installation as shown in Fig. 7.8.

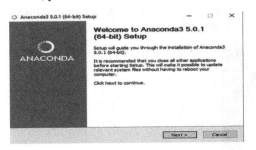

Fig. 7.8 Python version for Anaconda

Then agree to the license as in Fig. 7.9.

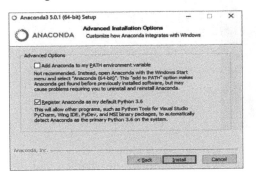

Fig. 7.9 License Agreement

At the Advanced Installation Options screen, I recommend that you do not check "Add Anaconda to my PATH environment variable" as shown in Fig. 7.10

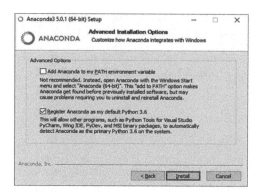

Fig. 7.10 Click Install for Anaconda

After the installation of Anaconda is complete, you can go to the Windows start menu and select the Anaconda Prompt as in Fig. 7.11

Fig. 7.11 Anaconda search in system

This opens the Anaconda Prompt. Anaconda is the Python distribution and the Anaconda Prompt is a command line shell (a program where you type in commands instead of using a mouse). The black screen and text that makes up the Anaconda Prompt doesn't look like much, but it is really helpful for problem solvers using Python.

At the Anaconda prompt, type python and hit [Enter]. The python command starts the Python interpreter, also called the Python REPL (for Read Evaluate Print Loop). you will able to access the python from terminal of anaconda prompt. From the anaconda prompt you can install all the libraries of python which are required in several projects to build.

On MacOS

Visit the Anaconda downloads page. Go to the following link: Anaconda.com/downloads

Select MacOS and download the.pkg installer

In the operating systems box, select [MacOS]. Then download the most recent Python 3 distribution (at the time of this writing the most recent version is Python 3.6) graphical installer by clicking the Download link. Python 2.7 is legacy Python. For problem solvers, select the most recent Python 3 version as in Fig. 7.12.

Fig. 7.12 Selecting version for Mac OS

Next Navigate to the Downloads folder and double-click the.pkg installer file you just downloaded. It may be helpful to order the contents of the Downloads folder by date to find the.pkg file.

Follow the installation instructions. It is advised that you install Anaconda for the current user and that Anaconda is added to your PATH.

Once Anaconda is installed, you need to load the changes to your PATH environment variable in the current terminal session.

Open the MacOS Terminal and type:

```
$ cd ~
$ source.bashrc
```

Open the MacOS Terminal and type:

```
$ python
```

You should see something like

```
Python 3.6.3 | Anaconda Inc. |
At the Python REPL (the Python >>> prompt) try:
>>> import this
```

If you see the Zen of Python, the installation was successful. Exit out of the Python REPL using the command exit (). Make sure to include the double parenthesis () after the exit command.

```
>>> exit ()
```

On Linux

Go to the following link: Anaconda.com/downloads. On the downloads page, select the Linux operating system as in Fig. 7.13.

Fig. 7.13 Linux version

In the Python 3.6 Version* box, right-click on the [64-Bit (x86) Installer] link. Select [copy link address] as shown in Fig. 7.14

Fig. 7.14 Copy the Downloading link of anaconda

Now that the bash installer (.sh file) link is stored on the clipboard, use wget to download the installer script. In a terminal, cd into the home directory and make a new directory called tmp. cd into tmp and use wget to download the installer. Although the installer is a bash script, it is still quite large and the download will not be immediate (Note the link below includes <release>. the specific release depends on when you download the installer).

```
$ cd ~
$ mkdir tmp
$ cd tmp
$ https://repo.continuum.io/archive/Anaconda3<release>.sh
```

With the bash installer script downloaded, run the.sh script to install Anaconda3. Ensure you are in the directory where the installer script downloaded:

```
$ ls
Anaconda3-5.2.0-Linux-x86 _ 64.sh
Run the installer script with bash.
$ bash Anaconda3-5.2.0-Linux-x86 _ 64.sh
```

Accept the Licence Agreement and allow Anaconda to be added to your PATH. By adding Anaconda to your PATH, the Anaconda distribution of Python will be called when you type $ python in a terminal.

Now that Anaconda3 is installed and Anaconda3 is added to our PATH, source the .bashrc file to load the new PATH environment variable into the current terminal session. Note the .bashrc file is in the home directory. You can see it with $ ls -a.

```
$ cd ~
$ source.bashrc
```

To verify the installation is complete, open Python from the command line:

```
$ python
Python 3.6.5 |Anaconda, Inc.|
[GCC 7.2.0] on Linux
Type "help", "copyright", "credits" or "license" for more
information.
>>>
```

If you see Python 3.6 from Anaconda listed, your installation is complete. To exit the Python REPL, type:

```
>>> exit ()
```

Using Jupyter

The simplest way to install Jupyter notebooks is to download and install the Anaconda distribution of Python. The Anaconda distribution of Python comes with Jupyter notebook included and no further installation steps are necessary.

Below are additional methods to install Jupyter notebooks if you are not using the Anaconda distribution of Python.

Installing Jupyter on Windows using the Anaconda Prompt

To install Jupyter on Windows, open the Anaconda Prompt and type:

```
> conda install jupyter
```

Type y for yes when prompted. Once Jupyter is installed, type the command below into the Anaconda Prompt to open the Jupyter notebook file browser and start using Jupyter notebooks.

```
> jupyter notebook
```

Installing Jupyter on MacOS

To install Jupyter on MacOS, open the MacOS terminal and type:

```
$ conda install jupyter
```

Type y for yes when prompted.

If conda is not installed, the Anaconda distribution of Python can be installed, which will install conda for use in the MacOS terminal.

Problems can crop up on MacOS when using the MacOS provided system version of Python. Python packages may not install on the system version of Python properly.

Moreover, packages which do install on the system version of Python may not run correctly. It is therefore recommended that MacOS users install the Anaconda distribution of Python or use homebrew to install a separate non-system version of Python.

To install a non-system version of Python with homebrew, key the following into the MacOS terminal. See the homebrew documentation at https://brew.sh.

```
$ brew install Python
```

After homebrew installs a non-system version of Python, pip can be used to install Jupyter.

```
$ pip install jupyter
```

Installing Jupyter on Linux

To install Jupyter on Linux, open a terminal and type:

```
$ conda install jupyter
```

Type y for yes when prompted.

Alternatively, if the Anaconda distribution of Python is not installed, one can use pip.

```
$ pip3 install jupyter
```

Using google Collab

Google Collaboratory is a free online cloud-based Jupyter notebook environment that allows us to train our machine learning and deep learning models on CPUs, GPUs, and TPUs.

Collab gives us 12 hours of continuous execution time. After that, the whole virtual machine is cleared and we have to start again. We can run multiple CPU, GPU, and TPU instances simultaneously, but our resources are shared between these instances.

Getting Started with Google Collab:

Go to this link and sign up or login with your google account

https://colab.research.google.com/as seen in Fig. 7.15

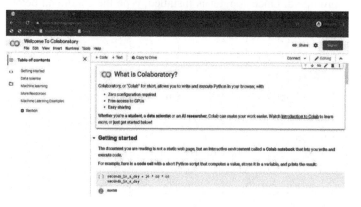

Fig. 7.15 Google Collaboratory

After login you will see this screen as in Fig. 7.16

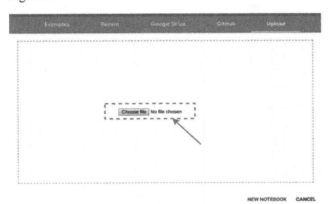

Fig. 7.16 Google Collaboratory Homepage

Click on the NEW NOTEBOOK button to create a new Collab notebook. You can also upload your local notebook to Collab by clicking the upload button as shown in Fig. 7.17:

Fig. 7.17 uploading Notebook

You can also import your notebook from Google Drive or GitHub, but they require an authentication process.

Fig. 7.18 Notebook Name change

As seen in Fig. 7.18 you can rename your notebook by clicking on the notebook name and change it to anything you want. I usually name them according to the project I'm working on.

Google Collab Runtimes – Choosing the GPU or TPU Option

The ability to choose different types of runtimes is what makes Collab so popular and powerful. Here are the steps to change the runtime of your notebook:

Click 'Runtime' on the top menu and select 'Change Runtime Type' as in Fig. 7.19

Fig. 7.19 Change Runtime Type

Here you can change the runtime according to your need as in Fig. 7.20:

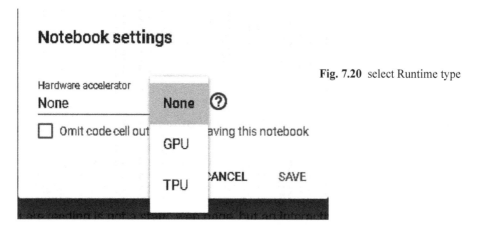

Fig. 7.20 select Runtime type

I implore you to shut down your notebook after you have completed your work so that others can use these resources because various users share them. You can terminate your notebook like this as in Fig. 7.21:

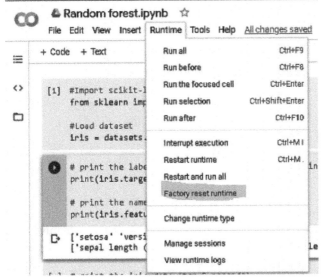

Fig. 7.21 Terminating Notebook

Fig. 7.21 Terminating Notebook

Using Terminal Commands on Google Collab

You can use the Collab cell for running terminal commands. Most of the popular libraries come installed by default on Google Collab. Yes, Python libraries like Pandas, NumPy, scikit-learn are all pre-installed.

If you want to run a different Python library, you can always install it inside your Collab notebook like this:

```
!pip install library _ name
```

Pretty easy, right? Everything is similar to how it works in a regular terminal. We just you have to put an exclamation (!) before writing each command like:

```
!ls
```

or:

```
!pwd
```

Uploading Files and Datasets:

Here's a must-know aspect for any data scientist. The ability to import your dataset into Collab is the first step in your data analysis journey.

The most basic approach is to upload your dataset to Collab directly as in Fig. 7.22:

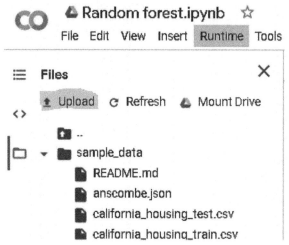

Fig. 7.22 Upload data

You can use this approach if your dataset or file is very small because the upload speed in this method is quite low. Another approach that I recommend is to upload your dataset to Google Drive and mount your drive on Collab. You can do this in just one click of your mouse as in Fig. 7.23:

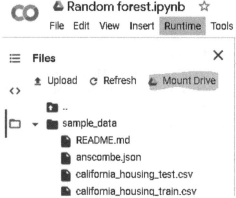

Fig. 7.23 Mounting Google drive

You can also upload your dataset to any other platform and access it using its link. I tend to go with the second approach more often than not (when feasible).

Saving Your Notebook:

All the notebooks on Colab are stored on your Google Drive. The best thing about Colab is that your notebook is automatically saved after a certain time period and you don't lose your progress.

If you want, from Fig. 7.24 you can export and save your notebook in both *.py and *.ipynb formats:

Fig. 7.24 Saving the Notebook

Not just that, you can also save a copy of your notebook directly as seen in Fig. 7.25 on GitHub, or you can create a GitHub Gist:

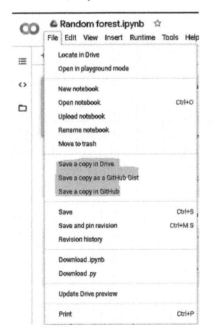

Fig. 7.25 Save Notebook

Sharing Your Notebook:

Google Collab also gives us an easy way of sharing our work with others. This is one of the best things about Collab as seen in Fig. 7.26:

Fig. 7.26 Sharing Notebook

Just click the Share button, and it gives us the option of creating a shareable link that we can share through any platform. As in Fig. 7.27 you can also invite others using their email IDs. It's exactly the same as sharing a Google Doc or Google Sheet. The intricacies and simplicity of Google's ecosystem are astounding!

Fig. 7.27: Link generating

Summary

Tools for building a project in Artificial Intelligence as discussed can be like python, jupyter, google Collaboratory. By studying all these tools we can clearly say that now we can easily work on these environment for biding a real time project which can be useful. Rather than all these you can go foe amazon EC2 server which will give you 1 year free trail to host any model or project.

Chapter 8

Important Libraries for AI

In this chapter we will cover all the basics libraries and some frameworks of machine learning, deep learning and computer vision. We will discuss all about their use with an example and instruction which will states the commands to install to those libraries for building any project in our system. So let's begin

What is a Library?

Library is an aggregate of features and methods that allow you to take many actions without getting your own code written.

For instance, the libraries that you have to know are if you work with data, numpy, scipy, pandas, etc. They have very simple data processing functions that save you time for small tricks.

Machine Learning and Python

- **TensorFlow:** Google, along with the team of Brain, created this library. For almost every Google machine learning program, TensorFlow is used. As neural networks can be easily represented as computer graphs, they can be used as a Tensor operation using the TensorFlow method, TensorFlow operates like a code-library for writing new algorithms involving a number of tensor operations. Moreover, tensors represent the data in N-dimensional matrices.

- Features of TensorFlow: For fast linear algebra operations, Tensor Flow is optimized, employs techniques like XLA.

  - **Responsive Construct:** With TensorFlow, each part of the graph that is not an option when using NumPy or SciKit can be visualized easily.

  - **Flexible:** One of the most important aspects of TensorFlow is that its operability is versatile, which means it has modularity and the parts of it you want to render independently.

  - **Easy Trainable:** On CPU and GPU for distributed computing it is easily trained.

  - **Parallel Neural Network Training:** TensorFlow provides pipelining to train multiple neural networks and multiple GPUs, making modeling on large-scale systems extremely effective.

  - **Large Community:** Naturally, a large team of software engineers has formed by Google that works continuously on stabilization enhancements.

  - **Opensource:** The great thing about this learning machine library is that it is open source so that anyone can use it when they want just by downloading it from online.

- Command to Install Scikit-learn

  Pip install tensorflow

  Example:

```
# Python program using TensorFlow
# for multiplying two arrays

# import `tensorflow`
import tensorflow as tf

# Initialize two constants
x1 = tf.constant ([1, 2, 3, 4])
x2 = tf.constant ([5, 6, 7, 8])

# Multiply
result = tf.multiply (x1, x2)

# Initialize the Session
sess = tf.Session ()

# Print the result
print (sess.run (result))

# Close the session
sess.close ()

Output:
[ 5 12 21 32]
```

- **Scikit-learn**: It is a Python library with NumPy and SciPy connections. It is recognized as one of the best libraries for complex data processing. It is built on Numpy, Scipy and Matplotlib.

  This library includes a number of improvements. One improvement is the cross-validation function that allows more than one metric to be used. Some training techniques such as regression of logistics and closest neighbors have been enhanced.

  - Features of Scikit-Learn

    Cross Validation: Specific techniques for the accuracy of tracked models on unsightly data are available

    Unsupervised Learning Algorithm: A broad range of algorithms – from clustering, factor analysis, main component analysis to unsupervised neural networks – are available in the package.

    Feature Extraction: Useful for extracting image and text features (e.g. word bag)

- Command to Install Scikit-learn

  Pip install scikit-learn

- **NumPy:** It is considered to be one of Python's most popular machine Learning library. For multiple operations on tensors, TensorFlow and other libraries internally use NumPy. NumPy's best and main feature is the Array interface.

  - Features of NumPy:

    Interactive: NumPy is very interactive and user-friendly.

    Mathematics: Makes complex math very simplistic.

    Intuitive: It is quick to code and it is easy to understand the concepts.

    Lot of Interaction: A lot of open source participation, therefore, it is widely used.

    Command to Install NumPy –

    Pip install numpy

    Example for NumPy: (Save the File as Num1.py)

```
# Python program using NumPy
# for some basic mathematical
# operations

import numpy as np

# Creating two arrays of rank 2
x = np.array ([[4, 5], [6, 8]])
y = np.array ([[8, 9], [10, 11]])

# Creating two arrays of rank 1
v = np.array ([12, 13])
w = np.array ([15, 16])

# Inner product of vectors
print (np.dot (v, w), "\n")

# Matrix and Vector product
print (np.dot (x, v), "\n")

# Matrix and matrix product
print (np.dot (x, y))

Output:
>>>
388

[113 176]
```

```
[[ 82  91]
 [128 142]]
>>>
```

- **SciPy:** It is an application creation and engineering machine learning library. Nonetheless, the gap between SciPy and SciPy stack still has to be identified. SciPy library contains optimization modules, linear algebra, integration modules and statistics.

Features of SciPy:

The key advantage of the SciPy library is that NumPy is used for the creation of the array.

In addition, with its unique substructures SciPy offers all the powerful numerical routines such as optimization, numerical integration and many others.

Command to Install scipy –

Pip install scipy

- **Pandas:** Pandas is a Machine Learning library in Python which offers high-level data structures and a wide range of analytical tools. The ability to translate complex operations with data using one or two commands is one of the major features of this library. Pandas have many standardized grouping methods, data combination and filtering methods, as well as functions for time series. Those are all accompanied by excellent speed measurements.

- Features of pandas: Pandas will make it easier to manipulate data throughout. Pandas' function highlights provide support for operations such as re-indexing, Iteration, Sorting, aggregations, concatenations and visualizations.

Command to Install pandas –

Pip install pandas

- **Matplotlib**: Matplotlib is a popular data visualization library for Python. It's not related to machine learning directly, like Pandas. It is especially useful to imagine patterns in the data if a programmer wants to. This is a library for 2D plotting used for the development of 2D plots. A module called pyplot makes it easy for plotting programmers to control line styles, fonts, axes, etc. It offers different types of graphs and plots for viewing data, e.g. histograms, error charts and chats, etc.

Command to Install Matplotlib –

Pip install matplotlib

Example

```
# Python program using Matplotib
# for forming a linear plot

# importing the necessary packages and modules
import matplotlib.pyplot as plt
import numpy as np
```

```
# Prepare the data
x = np.linspace (0, 10, 100)

# Plot the data
plt.plot (x, x, label ='linear')

# Add a legend
plt.legend ()

# Show the plot
plt.show ()
```

Output: as Fig. 8.1

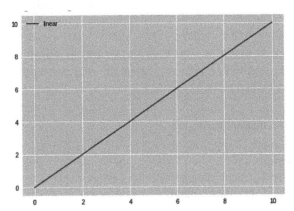

Fig. 8.1 Program output

- **Keras:** It is considered as one of the best machine learning libraries in Python. This offers an simpler way to describe neural networks. Keras provides some of the best tools for model creation, data-sets analysis, graph visualization and much more. Keras uses Theano or TensorFlow internally in the background. This may also use some of the best-known neural networks as CNTK. When comparing it with other learning machine libraries, Keras is comparatively sluggish. Since it uses back-end technology to generate a data graph and then uses it to perform operations. In Keras, all models are mobile.

Feature of Keras:

- On CPU and GPU operate smoothly.
- Keras supports virtually all neural network models-completely connected, convolutionary, pooling, recurrent, convergence, etc. Such models can also be combined to create more complex models.
- The modular design of Keras is incredibly articulate, versatile and perfect for creative work.

- Keras is a Python-based framework that makes debugging and exploring simple. Command to Install Keras –

Pip install Keras

- **PyTorch:** It is the biggest machine educational library which enables developers to perform tensor calculations by accelerating the GPU, creating dynamic graphs and automatically calculating gradients. However, PyTorch provides a wide range of APIs for the resolution of neural network device issues. This machine Learning library is developed in the form of Torch, an open source software library in C with a Lua wrapper. Released in 2017, this library in Python became popular and attracts a growing number of machine learning developers since its inception.

Features of PyTorch:

- Hybrid Front End: A new hybrid front end provides ease of use and versatility in the eager mode while transitional in C++ operating environments into a graphic mode for speed, optimization and functionality.
- Distributed Training: Optimize research and development efficiency by benefiting from native assistance to a synchronize collective operations and peer-to-peer networking accessible from Python and C++

Command to Install Pytorch –

Pip install torchvision

- **Theano**: It is a Python computational framework library for multidimensional array computing. TensorFlow functions similarly to Theano, though not as strong as TensorFlow. Due to its inability to adapt to manufacturing environments. Theano can be used in TensorFlow-like distributed or parallel environments

Features of Theano:

- Tight integration with NumPy: Capacity in Theano-compiled functions to use completely NumPy arrays.
- Transparent use of a GPU: Perform computations with data intensity much more quickly than a CPU.
- Efficient symbolic differentiation: For functions of one or more inputs, Theano does the derivatives.
- Speed and stability optimizations: Get the right log $(1+x)$ answer even if x is very small. It is only one example of Theano's stability.
- Dynamic C code generation: Evaluate terms faster than ever, thus making them much more effective.
- Extensive unit-testing and self-verification: Detect and diagnose many types of models of errors and ambiguities.

Command to Install Theano –

Pip install Theano

Deep Learning

- **Mxnet:** It is a deep learning open-source framework used for neural networking and training. It is highly productive and fast. It can scale through several GPUs and several machines.

To Install Maxnet – Download the package and install the package

https://mxnet.apache.org/get_started/?

Link for more details: https://mxnet.apache.org/

- Caffe: Caffe is a deep learning framework made with expression, speed, and modularity in mind. It is developed by Berkeley AI Research (BAIR) and by community contributors.

To Install Caffe– prefer the documentation on this link

https://caffe.berkeleyvision.org/installation.html

Link for more details: https://caffe.berkeleyvision.org/

- Fast.ai: The fastai library simplifies training fast and accurate neural nets using modern best practices. It's based on research in to deep learning best practices undertaken at fast.ai, including "out of the box" support for vision, text, tabular, and collab (collaborative filtering) models.

To Install Fast.ai– prefer the documentation on this link

https://github.com/fastai/fastai

Link for more details:https://docs.fast.ai/

- **CNTK:** A toolkit for commercial-grade distributed deep learning is the Microsoft Cognitive Toolkit (CNTK). Neural networks are defined via a guided chart as a series of computer measures. CNTK makes it simple to incorporate and combine common models such as feed-forward DNNs, CNNs and recurring neural networks (RNNs/ LSTMs). CNTK allows automatic differentiation, and parallelization, of stochastic gradient descent (SGD) learning through multiple GPUs and servers.

CNTK supports 64-bit Linux or 64-bit Windows operating systems. To install you can either choose pre-compiled binary packages, or compile the toolkit from the source provided in GitHub:https://github.com/Microsoft/CNTK

Link for more details:https://docs.microsoft.com/en-us/cognitive-toolkit/

ComputerVision

- **OpenCV**: OpenCV is one of the most widely used libraries for computer vision applications. Python API for OpenCV is the OpenCV Python. OpenCV-Python is not only easy, as the context of the C/C++ code is easily coded and deployed (as a result of Python wrapper). This makes the execution of computer-intensive vision programs a good choice.

Command to Install OpenCV–

Pip install opencv-python

- **Imutils:** The simple image processing tasks such as translation, rotate, resize, skeletonization, display, Matplotlib images, contours sorting and edges are simplified with OpenCV and Python in a variety of convenience features.

Command to Install imutils –

Pip install imutils

- **Pillow/PIL (python Imaging Library):** PIL (Python Imaging Library) is a free Python programming language library that supports several different image file formats for opening, handling, and saving. Nevertheless, with its last release in 2009, its growth has stagnated. Thankfully, Pillow, an popular PIL fork, is more easily installed, running on all major operating systems, and supporting Python 3. The library has simple image processing features, such as point operations, filtering with built-in cooling kernels and conversions of color space.

  Command to Install pillow –

  Pip install pillow

- **Scikit-image:** Scikit-image is the Python programming language open-source image processing library. The package includes segmentation algorithms, geometrical transformations, color space manipulation, visualization, filtering, morphology, function detection and much more. Scikit-image is a series of image processing algorithms. This is open and free of any restrictions. This requires some implementations of algorithms that OpenCV does not.

  Command to Install Scikit-image –

  Pip install Scikit-image

  Link for more details:https://scikit-image.org/

- **SimpleCV:** Another open source platform for computer vision applications is SimpleCV. It allows access to various powerful computer vision libraries like OpenCV without knowing bit depth, file formats, color spaces, and so on. The curve of his education is slightly smaller than that of OpenCV and (as his slogan says) "it's easy to see a machine."

  Link for more details:http://simplecv.org/

- **Vigra:** It is a computer vision library that focuses on scalable algorithms as the key know-how of algorithms in this area. As a result, the book was build in the standard template library, using generic programming pioneered by Stepanov and Musser. You can use VIGRA algorithms in addition to your data structures within your environment by writing a few adapters (image iterators and accessories). Also, you can use VIGRA data structures that are easy to adapt to a wide variety of applications. VIGRA has practically no versatility, as the architecture uses polymorphism (templates) for compile time

  Link for more details: https://ukoethe.github.io/vigra/

- **Mahotas:** It is another library for Python's computer vision and picture processing. It includes conventional image treatment functions, such as filters and morphological operations, as well as more modern machine viewing functions, including interest point detection and local descriptors for object computation. The Python code is ideal for rapid development, but the algorithms are used in C++ and speed tuned. The library of Mahotas is fast with minimal code and even minimum dependency.

Link for more details:https://mahotas.readthedocs.io/en/latest/

- **Pycairo**: Pycairo is a Python module that supplies the library with bindings. It works with Python 3.5 + and relies on cairo > = 1.13.1. Pycairo is authorised under the LGPL-2.1-only OR MPL-1.1 including this report. The Pycairo linking systems have been designed to suit the Cairo API as closely as possible, and only deviate if explicitly more 'pythonically' deployed.

Featrures of pycairo:

  - Offers a cairo-focused object gui.
  - Searches and converts error status of artifacts to exceptions.
  - The C API is provided so other Python extensions can be used.

Link for more details:https://pypi.org/project/pycairo/

Pgmagick: For the Graphics Magick library, the pgmagick is a Python-based wrapper. The image treatment system Graphics Magick is sometimes referred to as the Swiss Army Knife. The powerful and effective range of images in over 88 major formats, including DPX, GIF, JPEG, JPEG-2000, PNG, PDF, PNM and TIFF, facilitates the reading, writing and editing of photos.

Link for more details:https://pypi.org/project/pgmagick/

Summary

Here we have seen some of the important libraries of these domains of artificial intelligence and also have covered example for some of them with justify their use. There are so many libraries for python which can be useful for different projects to built in artificial intelligence and other area of study for different perspective and objective.

Chapter 9

Machine Learning Algorithms

In this chapter of machine learning algorithms we will cover the basic algorithms of machine learning with their implementation using dataset and discuss their pro's and cons of algorithms for better understanding and how to use them in any project or with any dataset to get an correct result which can useful for research or study.

Key Component

We defined a data set composed of audio, images and binary labels that provided an understanding of how we could train a model from snippets to classifications. This sort of problem is one of many kinds of machine learning problems when we try and predict a specified unknown label given known inputs, provided that a dataset of examples for which labels are known is called supervised learning. Let's discus what are the major key components we have to get there

- Data
- Models
- Algorithms

Linear Regression

Linear regression is a linear method in statistics to model the relationship between one or more independent variables and one (y) dependent variable (s). The relationships are modeled in linear regression with linear prediction functions which are estimated by unknown model parameters from the data. Linear regression is one of Machine Learning's most common algorithms. It is because of its relative simplicity and popular properties. It is two types which are

- **Simple Linear Regression**: If you only work with one independent variable, linear regression is called simple.

  It is Denoted by the formula $f(x) = mx + b$

  We will use two functions like cost and gradient to calculate and optimize our model with the data we provide.

  Cost Function: We calculate the precision with the mean squared error (squared) cost function of our linear regression algorithm. The average squared distance between the expected and the real performance is calculated by MSE.

  Its formula is

  $$\text{Error}(m, b) = \frac{1}{N} \sum_{i=1}^{N} (\text{Actual output} - \text{Predicted output})^2$$

MSE is very simple and can be easily coded with Python like

```
def cost_function (m, b, x, y):
    totalError = 0
    for i in range (0, len (x)):
        totalError += (y[i]- (m*x[i]+b))**2
    return totalError/float (len (x))
```

Optimization: We will use gradient descent to find the coefficients which minimize our error function. Gradient descent is an algorithm of optimization that iteratively takes steps towards the local cost function.

In order to find the way to the bottom, we have to take the error function in relation to our m and b-slope. Then we take a step in the negative direction.

General Gradient Descent Formula

$$\Theta_j = \Theta_j - \alpha \frac{\partial}{\partial \Theta_j} J(\Theta_0 - \Theta_1)$$

Gradient Descent for Simple Linear Regression.

$$\frac{\partial}{\partial m} = \frac{2}{N} \sum_{i=1}^{N} -x_i (y_i - (mx_i + b))$$

$$\frac{\partial}{\partial b} = \frac{2}{N} \sum_{i=1}^{N} -(y_i - (mx_i + b))$$

The gradient descent implementation is a little more complex, but can also be easily accomplished in pure Python

```
def gradient _ descent (b, m, x, y, learning _ rate, num _ iterations):
  N = float (len (x))
  for j in range (num _ iterations): # repeat for num _ iterations
      b _ gradient = 0
      m _ gradient = 0
      for i in range (0, len (x)):
          b _ gradient += - (2/N) * (y[i] - ( (m * x[i]) + b))
          m _ gradient += - (2/N) * x[i] * (y[i] - ( (m * x[i]) + b))
      b -= (learning _ rate * b _ gradient)
      m -= (learning _ rate * m _ gradient)
      if j%50==0:
        print ('error:', cost _ function (m, b, x, y))
  return [b, m]
```

Implementing Linear Regression

We must construct a dataset and identify several initial variables in order to run our linear regression model. For simple random and matplotlib data to be shown, we use NumPy.

```
import numpy as np
import matplotlib.pyplot as plt


x = np.linspace (0, 100, 50)
delta = np.random.uniform (-10, 10, x.size)
y = 0.6*x + 5 + delta
plt.scatter (x, y)
plt.show ()
```

Output: as in Fig. 9.1

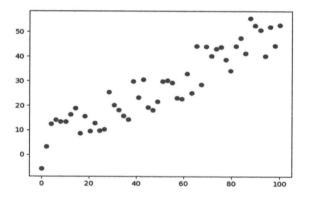

Fig. 9.1 Program Output

Now that we have our data, we are ready to use the above functionality to train our model:

```
# defining some variables
learning _ rate = 0.0001
initial _ b = 0
initial _ m = 0
num _ iterations= 100


print ('Initial error:', cost _ function (initial _ m,
                                  initial _ b, x, y))
[b, m] = gradient _ descent (initial _ b, initial _ m, x, y,
                         learning _ rate, num _ iterations)
print ('b:', b)
```

```
print ('m:', m)
print ('error:', cost _ function (m, b, x, y))
plt.show ()
```

Output:
```
Initial error: 1576.099063465165
error: 205.2155773301355
error: 41.73021903795362
b: 0.03851474971994741
m: 0.6744456473643745
error: 41.69053444198719
```

We can now use our model as a predictor, as we have trained it. We will only predict the training data in this simple example and use the findings to plot the best matplotlib line.

```
predictions = [ (m * x[i]) + b for i in range (len (x))]
plt.scatter (x, y)
plt.plot (x, predictions, color='r')
plt.show ()
```

Output: as in Fig. 9.2

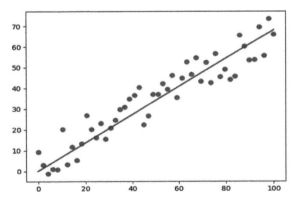

Fig. 9.2 Program Output

Multiple Linear Regression: When you deal with at least two independently operated variables, the linear regression is called multivariate. Each of the variables also known as characteristics is multiplied by a weight that our linear regression algorithm has learned.

It is Denoted by the formula

$$f(x) = b + W_1 X_1 + W_2 X_2 + \cdots W_n X_n = b + \sum_{i=0}^{n} W_i X_i$$

We will use two functions like cost and gradient to calculate and optimize our model with the data we provide.

Cost Function: We will use a mean squared error as a loss function, just as we did with a Simple linear regression. The only difference is that our prediction performance now comes from another feature.

As it is now possible to use NumPy to code our cost function with multiple features and weights:

Logistic Regression:

This is an algorithm of classification, used in categories of the response vary. The idea of the regression is to create a link between characteristics and the likelihood of a specific result.

For instance. The answer variable has two values, pass and fails when we need to predict if a student passes the test or fails if the number of hours studied is given as the factor.

This type of problem is called a binomial logistical regression, in which the answer variable has two 0 and 1 values or passes and fails or is true and false. The Multinomial Logistic Regression discusses situations where three or more potential values can be present in the response variable.

Why Logistic Regression, But not Linear?

Let 'x' be some function with binary classification, and let 'y' be the output that can be either 0 or 1.

In view of the input, the probability that the output is 1 is:

$$P(y = 1/x)$$

We may state that if we estimate the probability by means of linear regression:

$$p(X) = \beta_0 + \beta_1 X$$

where, $$p(x) = p(y = 1/x)$$

The predicted probability can be produced by a linear regression model as any number ranges from negative to positive, while an outcome may be only between $0 < P(x) < 1$ as in Fig. 9.3.

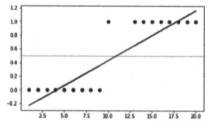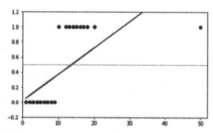

Fig. 9.3 Logistic Regression

Linear regression also has a huge effect on outliers.

To avoid this problem, log-odds function or logistic function is used

Logistic/logit Function

Logistic regression can be Denoted as:

$$\log\left(\frac{p(X)}{1-p(X)}\right) = \beta_0 + \beta_1 X$$

The logit or log-odds function is on the left side, and $p(x)/(1 - p(x))$ is referred to as odd.

The chances represent the ratio of success probability to loss probability. Therefore, a linear combination of inputs with the log (odds) in the Logistic Regression is mapped-output equal to 1.

If the above function is inversed then we get:

$$p(X) = \frac{e^{\beta_0 + \beta_1 X}}{1 + e^{\beta_0 + \beta_1 X}}$$

This equation is known to be Sigmoid function which is a curve that provides an S-shaped curve. It also provides a probability value of $0 < p < 1$.

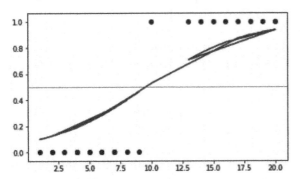

Fig. 9.4 Sigmoid Function

Estimation of Regression Coefficients

We use Maximum Likelihood Estimation, unlike linear regression, which uses Ordinary Last Square for parameter estimation.

There can be limitless collections of coefficients of regression. The maximums probability is that the maximum possibility of obtaining the data we have observed is a series of regression coefficients.

If we have binary results, it's only β if the result was good and $1 - \alpha$ otherwise. We have the probability function therefore:

$$\mathcal{L}(\beta; y) = \prod_{i=1}^{N}\left(\frac{\pi_i}{1-\pi_i}\right)^{y_i}(1-\pi_i)$$

To evaluate the parameter value, the probability function log is adopted because it does not modify the function properties.

The log-like nature is distinguished, and parameter values that maximize the log-likelihood are calculated using iterative techniques such as Newton.

Performance of Logistic Regression Model: Deviance is used instead of sum of squares to test the output of the logistic regression model as seen in table 9.1.

- Null Deviance demonstrates the expected reaction of a model with an interception only.
- System deviation indicates the answer to the addition of independent variables expected by a system. The parameter or set of parameters can be inferred that the model's variance is substantially smaller than the null deviance.
- The Confusion Matrix is another way to assess the accuracy of the model.

Table 9.1 Logistic Regression model Performance

	P' (Predicted)	n' (Predicted)
P (Actual)	True Positive	False Negative
n (Actual)	False Positive	True Negative

The Accuracy of the model is defined by:

$$\frac{\text{True Positives} + \text{True Negatives}}{\text{True Positive} + \text{True Negatives} + \text{Flase Positives} + \text{Flase Negatives}}$$

Multi-class Logistic Regression

Behind multi-class and binary regression, there is the same underlying intuition. However, we pursue a one v/s approach to multi-class issues.

For instance. We deal with a Multi-Class Problem if we have to predict if the weather is sunny, rainy or windy. We translate this into three question of binary classification, namely snowy and snowy, rainy or not, and windy or not. All three classifications are performed on feedback independently. The solution is the form for which the probability value is greater compared to others.

Implementation of Logistic Regression

We will simply implement logistic regression using Scikit-learn library which will allow us to use predefined logistic regression model. We will use iris Dataset:

The Iris dataset is perhaps one of the simplest machine learning datasets. The Iris dataset describes iris flowers in numerical form. It records sepal and petal length/width measurements can be seen in Fig. 9.5. Using these steps, we will try to predict flower species.

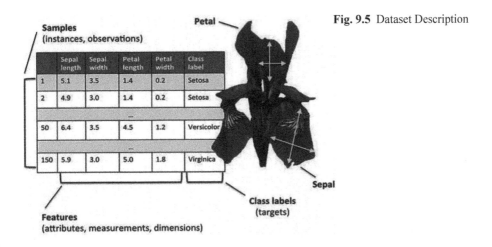

Fig. 9.5 Dataset Description

Which has columns regarding sepal length cm, sepal width cm, petal length cm, petal width cm, species in which first four Column are labels and last one is the target.

You can run these code in any python environment or Editor.

Step-1: we will import all the libraries required for implementation

```
# import dependencies
import numpy as np
import pandas as pd
import matplotlib.pyplot as plt
```

Note: The symbol # represent the comment for the program which meant for explanations of code

Step-2: loading the data

```
data = datasets.load_iris ()
```

here as we have imported the data through sckit's package, we will get the data in form of arrays,

```
# so here our labes is stored in
data.data
```

Output:

```
array ([[5.1,  3.5,  1.4,  0.2], [4.9, 3., 1.4, 0.2], [4.7, 3.2, 1.3,
0.2],[4.6, 3.1, 1.5, 0.2], [5., 3.6, 1.4, 0.2], [5.4, 3.9, 1.7, 0.4],[4.6,
3.4, 1.4, 0.3],[5., 3.4, 1.5, 0.2],[4.4, 2.9, 1.4, 0.2], [4.9, 3.1, 1.5,
0.1], [5.4, 3.7, 1.5, 0.2], [4.8, 3.4, 1.6, 0.2], [4.8, 3., 1.4, 0.1],
[4.3, 3., 1.1, 0.1], [5.8, 4., 1.2, 0.2], [5.7, 4.4, 1.5, 0.4], [5.4, 3.9,
1.3, 0.4], [5.1, 3.5, 1.4, 0.3], [5.7, 3.8, 1.7, 0.3],.......so on in this
way it will print all the data in the form of array on the screen.
# Targets for labels
data.target
```

Output:
```
array ([0, 0, 0, 0, 0, 0, 0, 0, 0, 0, 0, 0, 0, 0, 0, 0, 0, 0, 0, 0,
0, 0,0, 0, 0, 0, 0, 0, 0, 0, 0, 0, 0, 0, 0, 0, 0, 0, 0, 0, 0, 0, 0,
0, 0, 0, 0, 0, 0, 0, 1, 1, 1, 1, 1, 1, 1, 1, 1, 1, 1, 1, 1, 1, 1, 1,
1, 1, 1, 1, 1, 1, 1, 1, 1, 1, 1, 1, 1, 1, 1, 1, 1, 1, 1, 1, 1, 1,1,
1, 1, 1, 1, 1, 1, 1, 1, 1, 1, 1, 2, 2, 2, 2, 2, 2, 2, 2, 2, 2,2, 2,
2, 2, 2, 2, 2, 2, 2, 2, 2, 2, 2, 2, 2, 2, 2, 2, 2, 2, 2, 2, 2, 2,
2, 2, 2, 2, 2, 2, 2, 2, 2, 2, 2, 2, 2, 2, 2, 2])
```

Step 3: Now we will divide the data
we will divide the dataset into data and target for the model
```
X = data.data[:, :2] # X is the features in our dataset
y = (data.target != 0) * 1 # y is the Labels in our dataset
```

After separating we will visualize it in a graph using matplotlib
```
# here we have only taken the x line as SepalLengthCm and y line
as SepalWidthCm
plt.figure (figsize= (10, 6))
plt.scatter (X[y == 0][:, 0], X[y == 0][:, 1], color='g', label='0')
plt.scatter (X[y == 1][:, 0], X[y == 1][:, 1], color='r', label='1')
plt.legend ();
```

Output: As in Fig. 9.6

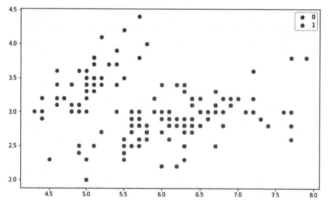

Fig. 9.6 Program Output

Step 4: Now will using scikit-learn logistic regression model to train and predict.
```
# import library with model
from sklearn.linear_model import LogisticRegression
# loading the model which c as parameter which can be given to
strengthen the model with predefined values like learning rate,
iteration etc.,
```

```
model = LogisticRegression (C=1e20)
# now we will train the model
model.fit (X, y)
```

Output:
```
LogisticRegression (C=1e+20, class _ weight=None, dual=False,
                    fit _ intercept=True,
    intercept _ scaling=1, l1 _ ratio=None, max _ iter=100,
    multi _ class='auto', n _ jobs=None, penalty='l2',
    random _ state=None, solver='lbfgs', tol=0.0001, verbose=0,
    warm _ start=False)
```

```
# Prediction from model
preds = model.predict (X)
 (preds == y).mean ()
```

Output:
```
1.0
```

Pros:
- Simple and efficient.
- Low variance.
- It provides probability score for observations.
- It can also be used for extraction of features

Cons:
- Does not handle several categorical characteristics/ attributes properly.
- It needs to transform non-linear characteristics.
- They are not sufficiently robust to handle more complex connections.

Decision Tree

The algorithm of the decision tree comes under the supervised learning category. A decision tree can be used to represent decisions and decision-making visually and clearly. This uses a tree-like decision pattern as the name goes. Although a commonly used data mining technique for a strategy to achieve a certain goal, it is often used extensively for machine learning.

Decision trees are designed using algorithms that recognize ways to break a collection of data based on various criteria. Classification trees are called tree models in which the target variable will take a distinct set of values. Decision trees where continuous values (typically genuine numbers) can be taken from the target variable are called regression trees. The general term for this is Classification and Regression Tree (CART).

Important Terminology in Decision Tree

Let us discuss the fundamental terms used in decision trees which can be seen in Fig. 9.7.

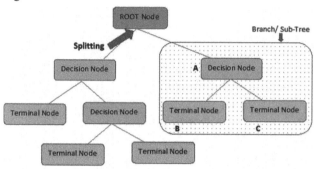

Fig. 9.7 Decision tree representation

Note:- A is parent node of B and C.

- **Root node:** It represents a whole population or sample and is further split up into two or more uniform samples.
- **Splitting:** This is a mechanism by which a node is split into two or more sub nodes.
- **Decision Node:** If a sub node splits into additional sub nodes, it is referred to as a decision node.
- **Terminal Node/Leaf:** Nodes without children (no further division) are referred to as Leaf nodes and Terminal nodes.
- **Pruning:** when we reduce the size of decision trees by eliminating nodes (the opposite to splitting).
- **Branch/Sub-Tree:** It is referred to as the sub-section of decision-tree.
- **Parent and Child Node:** A node divided into sub-nodes is called the parent node of sub-nodes where the child of a parent node is the sub-node.

Decision Tree Making Approach

The biggest challenge in Decision Tree is to classify the attribute for each level of the root node. This method is referred to as selection of attributes. We have two common selection measures for attributes:

- **Information Gain:** When we use a node in a decision node, the training instances are split into smaller entropy sub-sets. This shift in entropy is determined by the gain of information. We ask various types of questions in every node of the tree during the decision process. On the basis of the question asked, we will measure the data gain.

  Definition: Suppose S is a set of instances, A is an attribute, S_v is the subset of S with $A = v$, and Values (A) is the set of all possible values of A, the

$$\text{Gain}(S, A) = \text{Entropy}(S) - \sum_{v \in} \text{Values}(A) \frac{|S_v|}{S} \cdot \text{Entropy}(S_v)$$

Entropy: Entropy is the uncertainty factor for a random variable and it describe the impurity of arbitrary instances. The greater the entropy, the greater the value of knowledge.

- **Gini Index:** The Gini index is a measure of how frequently a randomly selected variable will be defined incorrectly. This means that you would choose an attribute with a lower Gini index. Sklearn follows "Gini" index parameters and takes "gini" value by chance. The Formula is given below to calculate the Gini Index.

$$\text{Gini Index} = 1 - \sum_{j} p_j^2$$

Implementation of Decision Tree

Let's take a look at how a decision tree classification in Python can be implemented. To continue,

Step-1: we import the libraries.

```
from sklearn.datasets import load_iris
from sklearn.tree import DecisionTreeClassifier
from sklearn.model_selection import train_test_split
from sklearn.metrics import confusion_matrix
from sklearn.tree import export_graphviz
from sklearn.externals.six import StringIO
from IPython.display import Image
from pydot import graph_from_dot_data
import pandas as pd
import numpy as np
```

Step-2: Loading the dataset

```
iris = load_iris ()
X = pd.DataFrame (iris.data, columns=iris.feature_names)
y = pd.Categorical.from_codes (iris.target, iris.target_names)
X.head ()
```

Output:

	sepal length (cm)	sepal width (cm)	petal length (cm)	petal width (cm)
0	5.1	3.5	1.4	0.2
1	4.9	3.0	1.4	0.2
2	4.7	3.2	1.3	0.2
3	4.6	3.1	1.5	0.2
4	5.0	3.6	1.4	0.2

Step-3: Splitting the data

```
y = pd.get_dummies (y)
X_train, X_test, y_train, y_test = train_test_split (
    X, y, random_state=1)
```

Step-4: Using the decision tree classifier to classify the flowers

```
dt = DecisionTreeClassifier ()
dt.fit (X_train, y_train)
```

Output:
```
DecisionTreeClassifier (ccp_alpha=0.0, class_weight=None,
    criterion='gini',
max_depth=None, max_features=None, max_leaf_nodes=None,
min_impurity_decrease=0.0, min_impurity_split=None,
min_samples_leaf=1, min_samples_split=2,
min_weight_fraction_leaf=0.0, presort='deprecated',
random_state=None, splitter='best')
```

Step-5: Compiling and visualizing the decision tree
```
dot_data = StringIO ()
export_graphviz (dt, out_file=dot_data,
        feature_names=iris.feature_names)
(graph,) = graph_from_dot_data (dot_data.getvalue ())
Image (graph.create_png ())
```
Note how it includes Gini impurities, the total sample number, the criteria for classification and the number of left/right samples.

Output:

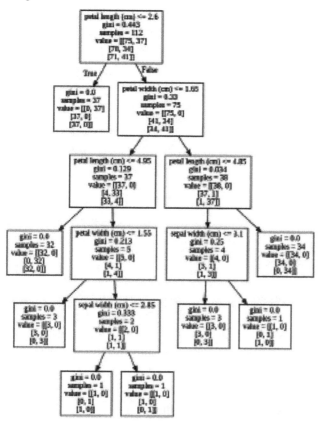

Step-6: Prediction with confusion matrix

```
y _ pred = dt.predict (X _ test)
species = np.array (y _ test).argmax (axis=1)
predictions = np.array (y _ pred).argmax (axis=1)
confusion _ matrix (species, predictions)
```

Output:
```
array ([[13,  0,  0],
[ 0, 15,  1],
[ 0,  0,  9]])
```

We can see that our decision tree listed 37/38 plants correctly.

Support Vector Machine

Support Vector Machine (SVM) is a categorized and regression analysis machine learning approach. It relies on supervised models of learning and training by algorithms. The vast volumes of data are analyzed for trends.

An SVM produces two parallel lines for parallel partitions. For each data category, it utilizes almost every attribute in a high-dimensional field. It separates the space for flat and linear partitions in a single pass. Divide the 2 groups into the clearest possible distance. Do so by a plane known as a hyperplane.

An SVM generates hyperplane with the highest margin to divide data into classes in a high-dimensional area. The distance between the two classes is the highest distance between such classes' nearest data points.

The bigger the margin, the lower the classifier classification generalization error is as in Fig. 9.8:

Fig. 9.8 SVM Representation

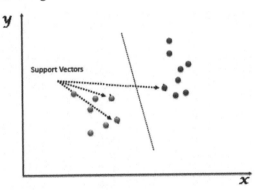

How does it work?

Consider two conditions in order to grasp the SVM algorithm:

- Separate case – Infinite boundaries can be split into two groups.
- Non-separable case – There is no division between two groups, but one overlap.

Implementation of SVM

Step 1: Importing the libraries

```
import numpy as np
import pandas as pd
import matplotlib.pyplot as plt
import seaborn as sns
from sklearn.svm import SVC
from sklearn.model_selection import train_test_split
from sklearn.metrics import classification_report,
    confusion_matrix
```

Step 2: importing and Using the dataset from google drive

```
from google.colab import drive
drive.mount ('/gdrive')
%cd /gdrive
```

Output:

Go to this URL in a browser: https://accounts.google.com/o/oauth2/auth?client_id=94.................y

Enter your authorization code:

....

Mounted at /gdrive

This is the code for mounting your drive with google research Collaboratory which will ask you authentication which you can do it with your Gmail account by just following the simple steps. As you can see in the output how Collaboratory will ask you to verify it by clinking in the given link then it will get mounted by giving a authentication to it.

```
#if you are using python editor just provide the file path which
is where you have stored on your system
# Import the dataset using Seaborn library
iris=pd.read_csv ('/gdrive/My Drive/Colab Notebooks/IRIS.csv')
#providing path for dataset

# Checking the dataset
iris.head ()
```

output:

	sepal_length	sepal_width	petal_length	petal_width	species
0	5.1	3.5	1.4	0.2	Iris-setosa
1	4.9	3.0	1.4	0.2	Iris-setosa
2	4.7	3.2	1.3	0.2	Iris-setosa
3	4.6	3.1	1.5	0.2	Iris-setosa
4	5.0	3.6	1.4	0.2	Iris-setosa

Step 3: Visualizing the similarities into the dataset using pair plots
```
# Creating a pairplot to visualize the similarities and especially
difference between the species
sns.pairplot (data=iris, hue='species', palette='Set2')
```

Output: as in Fig. 9.9

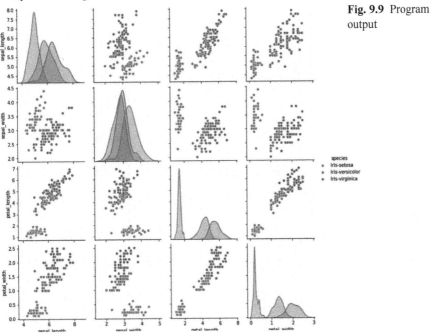

Fig. 9.9 Program output

Step 4: Splitting the test and train data
```
# Separating the independent variables from dependent variables
x=iris.iloc[:,:-1]
y=iris.iloc[:,4]
x _ train,x _ test, y _ train, y _ test=train _ test _ split
   (x,y,test _ size=0.30)
```

Step 5: Loading model and training

```
model=SVC ()
model.fit (x _ train, y _ train)
```

Output:

```
SVC (C=1.0, break _ ties=False, cache _ size=200,
    class _ weight=None, coef0=0.0, decision _ function _
    shape='ovr', degree=3, gamma='scale', kernel='rbf',
    max _ iter=-1, probability=False, random _ state=None,
    shrinking=True,tol=0.001, verbose=False)
```

Step 6: Results of prediction

```
pred=model.predict (x _ test)
# Importing the classification report and confusion matrix
print (confusion _ matrix (y _ test,pred))
```

Output:

```
[[15  0  0]
 [ 0 14  0]
 [ 0  2 14]]
#generating Classification report

print (classification _ report (y _ test, pred))
```

	precision	recall	f1-score	support
Iris-setosa	1.00	1.00	1.00	15
Iris-versicolor	0.88	1.00	0.93	14
Iris-virginica	1.00	0.88	0.93	16
accuracy			0.96	45
macro avg	0.96	0.96	0.96	45
weighted avg	0.96	0.96	0.96	45

Naïve Bayes

Naive Bayes algorithm can be described as a supervised Classification algorithm based on Bayes theorem with independence between features. It is also a technique of classification.

Bayes theorem is a principle for constructing classifiers behind this classification technique. The predictors are expected to be independent. In simple terms, the inclusion in a class of a particular characteristic is not connected to that of any other characteristic. The following is the Bayes theorem equation:

$$P(A|B) = P(B|A) \ P(A)/P(B)$$

Where

- $P(A|B)$ is the probability of hypothesis A given the data B. This is called the posterior probability.
- $P(B|A)$ is the probability of data B given that the hypothesis A was true.

- $P(A)$ is the probability of hypothesis A being true (regardless of the data). This is called the prior probability of A.
- $P(B)$ is the probability of the data (regardless of the hypothesis).

Implementation

Python libraries provide three types of classifiers of Naïve Bayes:

- **Gaussian:**. It assumes that the characteristics are distributed accordingly.
- **Multinomial:** For discrete counts,
- **Bernoulli:** If the function vectors are binary, the binomial model is useful (e.g. zeros and vectors)

In implementation we will be Gaussian naïve Bayes classifier and iris dataset. We already know the iris dataset as we have studied about it in before algorithms.

Of course, in this example only one and two classes are listed. However, also with a vector with many features and groups the process is the same (the only exception being the use of an Argmax function to return the most probable).

Step 1: Importing all libraries

```
from sklearn import datasets
import matplotlib.pyplot as plt
import pandas as pd
#importing the necessary packages
from sklearn.model_selection import train_test_split
from sklearn.naive_bayes import GaussianNB
```

Step 2: Dataset loading and Splitting

```
#downloading the iris dataset, splitting it into train set and
validation set
iris = datasets.load_iris ()
class_names = iris.target_names
iris_df=pd.DataFrame (iris.data, columns=iris.feature_names)
iris_df['target']=iris.target
iris_df.head ()
```

Output:

	sepal length (cm)	sepal width (cm)	petal length (cm)	petal width (cm)	target
0	5.1	3.5	1.4	0.2	0
1	4.9	3.0	1.4	0.2	0
2	4.7	3.2	1.3	0.2	0
3	4.6	3.1	1.5	0.2	0
4	5.0	3.6	1.4	0.2	0

```
iris_df.describe ()
```

Output:

	sepal length (cm)	sepal width (cm)	petal length (cm)	petal width (cm)	target
count	150.000000	150.000000	150.000000	150.000000	150.000000
mean	5.843333	3.057333	3.758000	1.199333	1.000000
std	0.828066	0.435866	1.765298	0.762238	0.819232
min	4.300000	2.000000	1.000000	0.100000	0.000000
25%	5.100000	2.800000	1.600000	0.300000	0.000000
50%	5.800000	3.000000	4.350000	1.300000	1.000000
75%	6.400000	3.300000	5.100000	1.800000	2.000000
max	7.900000	4.400000	6.900000	2.500000	2.000000

We have 4 features and a categorical goal because we have 3 levels (Setosa, Versicolor and Virginica, 0, 1 and 2). We can now build our classifiers accordingly:

```
# Now Splitting the data into test and train
X _ train, X _ test, y _ train, y _ test = train _ test _ split (iris _
df[['sepal length (cm)', 'sepal width (cm)', 'petal length (cm)',
'petal width (cm)']], iris _ df['target'], random _ state=0)
```

Step 3: Loading Model and compilation

```
NB = GaussianNB ()
NB.fit (X _ train, y _ train)
y _ predict = NB.predict (X _ test)
print ("Accuracy Naive Bayes: {:.2f}".format (NB.score (X _ test,
                                                        y _ test)))
```

Output:

```
Accuracy Naive Bayes: 1.00
```

As you can see, all test data are categorized correctly by our algorithm. It's a nice performance, the best one we will ever achieve. But take into account our weak dataset dimension (only 150 observations).Standard precision of 100 percent on real data is difficult to achieve. In addition, the argument for an error equal to zero may also be counterproductive. Indeed we prefer to have an algorithm that can generalize and not conform perfectly to test data, because the last goal of any ML algorithm is to make accurate predictions of new unknown data. The risk of overruns will lead to an unsustainable model if new data cannot be used (using too many parameters to create a model that perfectly fits test data).

Pros:

- Easy to comprehend.
- It can be trained on small data sets as well.

Cons:

- There are many sample corrections to solve this problem, such as "Laplacian correction," for features that have a null-frequency, the total likelihood is zero.
- Another drawback is that it very strongly assumes independence class features. These data sets in real life are almost difficult to locate.

K-Nearest Neighbors (KNN)

It is used to identify problems and to rectify them. It is used commonly to solve problems with classification. The key principle of this algorithm is that it stores all the existing cases and by voting plurality of its neighbors classifies new cases. The case is then assigned to the class which, according to a distance function, is the most common among his nearest K neighbors. The gap can be from Minkowski, Manhattan, Euclidean, and Hamming. Consider KNN as follows:

- Computationally KNN are expensive than other algorithms used for classification problems.
- The normalization of variables needed otherwise higher range variables can bias it.
- At KNN, we will work on a stage such as noise reduction, pre-processing.

Implementation

Step 1: Importing Libraries

```
import numpy as np
import pandas as pd
import matplotlib.pyplot as plt
from collections import Counter
```

Step 2: Loading And visualizing the data

```
names = ['sepal_length', 'sepal_width', 'petal_length',
'petal_width', 'class']
# Link for downloading dataset: https://archive.ics.uci.edu/ml/
machine-learning-databases/iris/
df = pd.read_csv ('C:/Users/EETRO/Desktop/Book Code/iris.data.
txt', header = None, names = names) #path where file is stored
at desktop
df.head ()
```

Output:

	sepal_length	sepal_width	petal_length	petal_width	class
0	5.1	3.5	1.4	0.2	Iris-setosa
1	4.9	3.0	1.4	0.2	Iris-setosa
2	4.7	3.2	1.3	0.2	Iris-setosa
3	4.6	3.1	1.5	0.2	Iris-setosa
4	5.0	3.6	1.4	0.2	Iris-setosa

```
setosa = df[df['class'] == 'Iris-setosa']
versicolor = df[df['class'] == 'Iris-versicolor']
virginica = df[df['class'] == 'Iris-virginica']

plt.plot (setosa['sepal_length'], setosa['sepal_width'], 'ro',
label = 'setosa')
```

```
plt.plot (versicolor['sepal_length'], versicolor['sepal_width'],
'bo', label = 'versicolor')
plt.plot (virginica['sepal_length'], virginica['sepal_width'],
'go', label = 'virginica')
plt.xlabel ('sepal_length')
plt.ylabel ('sepal_width')
plt.legend ()
plt.show ()
```

Output: as in Fig. 9.10

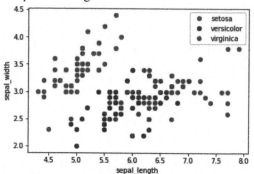

Fig. 9.10 Program Outcome

```
plt.plot (setosa['petal_length'], setosa['petal_width'], 'ro',
label = 'setosa')
plt.plot (versicolor['petal_length'], versicolor['petal_width'],
'bo', label = 'versicolor')
plt.plot (virginica['petal_length'], virginica['petal_width'],
'go', label = 'virginica')
plt.xlabel ('petal_length')
plt.ylabel ('petal_width')
plt.legend ()
plt.show ()
```

Output: as in Fig. 9.11

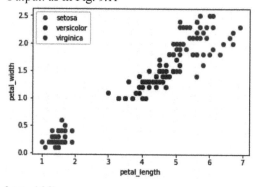

Fig. 9.11 Program outcome

```
len (df)
```

Output:
```
150
print ('length of setosa: %d' % (len (setosa)))
print ('length of versicolor: %d' % (len (versicolor)))
print ('length of virginica: %d' % (len (virginica)))
```

Output:
```
length of setosa: 50
length of versicolor: 50
length of virginica: 50
```

Step 3: Dividing the data into training and testing set with 3:1 ratio
```
df['is _ train'] = np.random.uniform (0, 1, len (df)) <=.75
df.head ()
```

Output:

	sepal_length	sepal_width	petal_length	petal_width	class	is_train
0	5.1	3.5	1.4	0.2	Iris-setosa	True
1	4.9	3.0	1.4	0.2	Iris-setosa	True
2	4.7	3.2	1.3	0.2	Iris-setosa	False
3	4.8	3.1	1.5	0.2	Iris-setosa	True
4	5.0	3.6	1.4	0.2	Iris-setosa	True

```
train, test = df[df['is _ train'] == True],
    df[df['is _ train'] == False]
print ('length of train dataset: %d' % (len (train)))
print ('length of test dataset: %d' % (len (test)))
```

Output:
```
length of train dataset: 111
length of test dataset: 39
```

```
train.columns
```

Output:
```
Index (['sepal _ length', 'sepal _ width', 'petal _ length',
'petal _ width', 'class', 'is _ train'], dtype='object')
```

```
train _ x = train[train.columns[:len (train.columns) - 2]]
train _ y = train['class']
test _ x = test[test.columns[:len (test.columns) - 2]]
test _ y = test['class']
```

Step 4: Calculation of distances by using function method
i.e., Euclidean Distance and Manhattan Distance

$$\text{Euclidean Distance} = \sqrt{\sum_{i=1}^{k}(x_i - y_i)^2}$$

$$\text{Manhattan Distance} = \sum_{i=1}^{k}|x_i - y_i|$$

```
#Euclidean Distance function takes inputs as
#data1 : first data point
#data2 : second data point

#and gives output as Euclidean distance

def euclidean _ distance (data1, data2):
    distance = 0
    for i in range (data2.shape[0]):
      distance += np.square (data1[i] - data2[i])
    return np.sqrt (distance)

#Manhattan Distance function takes inputs as
#data1 : first data point
#data2 : second data point

#and gives output as Manhattan distance

def manhattan _ distance (data1, data2):
    distance = 0
    for i in range (data2.shape[0]):
      distance += abs (data2[i] - data1[i])


    return distance
```

Step 5: Forming KNN function
Now KNN function takes inputs as

train_x : training samples,

train_y : corresponding labels,

dis_func : a function which calculates distance

sample : one test sample

k : number of training example to look for deciding the class of the given sample.

```
def knn (train _ x, train _ y, dis _ func, sample, k):
    distances = {}
    for i in range (len (train _ x)):
```

```
      d = dis_func (sample, train_x.iloc[i])
      distances[i] = d

sorted_dist = sorted (distances.items (),
                key = lambda x : (x[1], x[0]))

# take k nearest neighbors
   neighbors = []
   for i in range (k):
     neighbors.append (sorted_dist[i][0])
   # convert indices into classes
   classes = [train_y.iloc[c] for c in neighbors]

   # count each classes in top k
   counts = Counter (classes)

   # take vote of max number of samples of a class
list_values = list (counts.values ())
   list_keys = list (counts.keys ())
   cl = list_keys[list_values.index (max (list_values))]

   return cl #class of the sample

sl = knn (train_x, train_y, manhattan_distance,
          test_x.iloc[3], 5)
test_y.iloc[3]
```

Output:
'Iris-setosa'

Step 6: Getting the Accuracy of the models
```
def get_accuracy (test_x, test_y, train_x, train_y, k):
correct = 0
for i in range (len (test_x)):
sample = test_x.iloc[i]
true_label = test_y.iloc[i]
predicted_label_euclidean = knn (train_x, train_y,
euclidean_distance, sample, k)
if predicted_label_euclidean == true_label:
correct += 1
accuracy_euclidean = (correct/len (test_x)) * 100

correct = 0 # reset correct value to 0
for i in range (len (test_x)):
sample = test_x.iloc[i]
true_label = test_y.iloc[i]
predicted_label_manhattan = knn (train_x, train_y,
manhattan_distance, sample, k)
```

```
if predicted _ label _ manhattan == true _ label:
correct += 1
accuracy _ manhatten = (correct/len (test _ x)) * 100
print ("model accuracy with euclidean is %0.2f"
    % (accuracy _ euclidean))
print ("model accuracy with manhattan is %0.2f"
    % (accuracy _ manhatten))
get _ accuracy (test _ x, test _ y, train _ x, train _ y, 5)
```

Output:

Model accuracy with euclidean is 94.87

model accuracy with manhattan is 94.87

Accuracy is the ratio of the number of instances correctly identified to the number of instances. Here, we can see that the Euclidean and Manhattan range as a measure of similarity is 94.87 percent accurate.

Pros:

• No data assumptions.
• Simple to understand algorithm.
• Can be used for regression and classification.

Cons:

• High memory demand — In order to measure nearest K neighbors, all training data should be in the memory.
• Related features sensitive.
• Sensitive to the size of data as the distance to the nearest K points is computed.

K-means Clustering

It is used to resolve the problem of clusters, as the name implies. This is a kind of unsupervised Learning. The principal principle of the K-Means algorithm is to assign the data set across different clusters. The K-means cluster algorithm is used to identify and find patterns and to make better decisions for groups that are not clearly labeled in the data. The new data can be easily allocated to the most appropriate group once an algorithm is performed and groups identified.

In order to start with k-means, you must initialize the cluster centroides (K) by random means. K-means is an iterative, two-step algorithm as in Fig. 9.12:

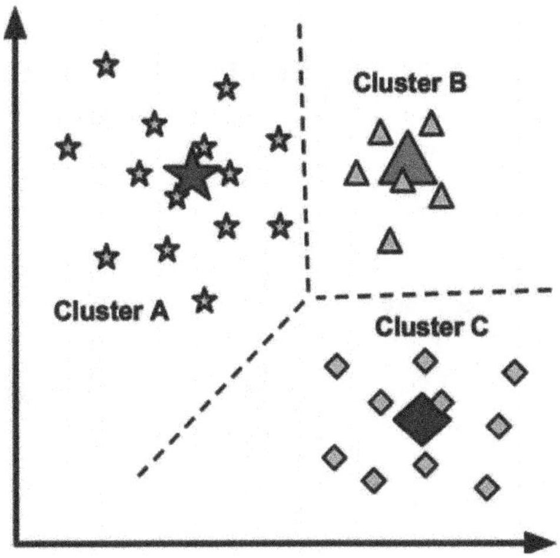

Fig. 9.12 Data Points Clustering

1. Assignment of the cluster: The algorithm traverses every data point and assigns data points to one of the 3 cluster centroids, depending on the cluster that is closer.
2. Move to centroid: Centroids are transferred to the mean of the clusters formed after the cluster assignment. And then it repeats the cycle. The centroids can no longer move around after a certain number of steps and then the iterations can end.

Implementation

Step 1: Importing of Libraries

```
from sklearn import datasets
import matplotlib.pyplot as plt
import pandas as pd
from sklearn.cluster import KMeans
```

Step 2: Loading the data
```
iris = datasets.load _ iris ()
```

Step 3: Defining the target and Predictors
$X = iris.data[:, :2]$

$y = iris.target$

Step 4: Visualizing the data
```
plt.scatter (X[:,0], X[:,1], c=y, cmap='gist _ rainbow')
plt.xlabel ('Speal Length', fontsize=18)
plt.ylabel ('Sepal Width', fontsize=18)
```

Output: as in Fig. 9.13

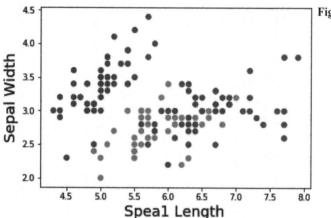

Fig. 9.13 Program outcome

Step 5: Loading and Compiling the model
```
km = KMeans (n _ clusters = 3, n _ jobs = 4, random _ state=21)
km.fit (X)
```

Output:
```
KMeans (algorithm='auto', copy _ x=True, init='k-means++',
max _ iter=300, n _ clusters=3, n _ init=10, n _ jobs=4,
precompute _ distances='auto', random _ state=21, tol=0.0001,
verbose=0)
```

Step 6: Getting Centroid
```
centers = km.cluster _ centers _
print (centers)
```

Output:
```
[[5.77358491 2.69245283]
[5.006 3.428 ]
[6.81276596 3.07446809]]
```

Step 7: Comparing the original data and clustered data
```
#this will tell us to which cluster does the data observations
belong.
new _ labels = km.labels _
# Plot the identified clusters and compare with the answers
fig, axes = plt.subplots (1, 2, figsize= (16,8))
axes[0].scatter (X[:, 0], X[:, 1], c=y, cmap='gist _ rainbow',
edgecolor='k', s=150)
axes[1].scatter (X[:, 0], X[:, 1], c=new _ labels, cmap='jet',
```

```
edgecolor='k', s=150)
axes[0].set _ xlabel ('Sepal length', fontsize=18)
axes[0].set _ ylabel ('Sepal width', fontsize=18)
axes[1].set _ xlabel ('Sepal length', fontsize=18)
axes[1].set _ ylabel ('Sepal width', fontsize=18)
axes[0].tick _ params (direction='in', length=10, width=5,
                       colors='k', labelsize=20)
axes[1].tick _ params (direction='in', length=10, width=5,
                       colors='k', labelsize=20)
axes[0].set _ title ('Actual', fontsize=18)
axes[1].set _ title ('Predicted', fontsize=18)
```

Output: as in Fig. 9.14

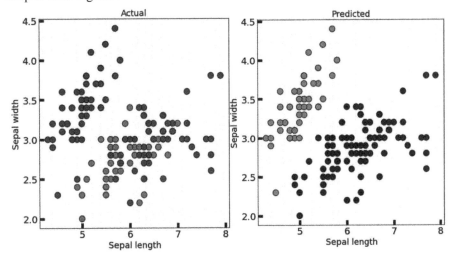

Fig. 9.14 Program outcome

Pros:

- K means are straightforward and machine efficient.
- It is highly intuitive and quick to imagine the effects.

Cons:

- K-means are strongly reliant upon scale and are not designed for data of different densities and shapes.
- More subjective is the estimation of performance. It needs much more human evaluation than metrics of confidence.

Random Forest

This is a supervised algorithm of classification. The random forest algorithm has the advantage that it could be used for problems of classification as well as regression. This is essentially the set of decision-making trees (forests) or the decision-making trees ensemble. The basic principle of the random forest is that each trees are graded and that the forest selects the best grades.

The distinction between random forest and decision tree algorithm is that the process of identifying the root node and dividing the function nodes runs randomly in random forests.

Implementation

Step 1: Importing all required libraries

```
#Import Random Forest Model
from sklearn.ensemble import RandomForestClassifier
#Import scikit-learn dataset library
from sklearn.model _ selection import train _ test _ split
from sklearn import datasets
import pandas as pd
#Import scikit-learn metrics module for accuracy calculation
from sklearn import metrics
#Import scikit-learn metrics module for accuracy calculation
from sklearn import metrics
```

Step 2: Loading and Visualizing the data

```
#Load dataset
iris = datasets.load _ iris ()


# print the label species (setosa, versicolor,virginica)
print (iris.target _ names)


# print the names of the four features
print (iris.feature _ names)
```

Output:

```
['setosa' 'versicolor' 'virginica']
['sepal length (cm)', 'sepal width (cm)', 'petal length (cm)',
'petal width (cm)']

# print the iris data (top 5 records)
print (iris.data[0:5])


# print the iris labels (0:setosa, 1:versicolor, 2:virginica)
print (iris.target)
```

Output:

```
[[5.1 3.5 1.4 0.2]
[4.9 3.  1.4 0.2]
[4.7 3.2 1.3 0.2]
[4.6 3.1 1.5 0.2]
[5.  3.6 1.4 0.2]]
[0 0 0 0 0 0 0 0 0 0 0 0 0 0 0 0 0 0 0 0 0 0 0 0 0 0 0 0 0 0 0 0 0 0 0 0
0 0 0 0 0 0]
```

```
0 0 0 0 0 0 0 0 0 0 0 0 0 0 1 1 1 1 1 1 1 1 1 1 1 1 1 1 1 1 1 1 1
1 1 1 1 1
1 1 1 1 1 1 1 1 1 1 1 1 1 1 1 1 1 1 1 1 1 1 1 1 1 1 2 2 2 2 2 2 2
2 2 2 2 2
2 2 2 2 2 2 2 2 2 2 2 2 2 2 2 2 2 2 2 2 2 2 2 2 2 2 2 2 2 2 2 2 2
2 2 2 2 2 2 2]
```

```python
# Creating a DataFrame of given iris dataset.
data=pd.DataFrame ({
'sepal length':iris.data[:,0],
'sepal width':iris.data[:,1],
'petal length':iris.data[:,2],
'petal width':iris.data[:,3],
'species':iris.target
})
data.head ()
```

Output:

	sepal length	sepal width	petal length	petal width	species
0	5.1	3.5	1.4	0.2	0
1	4.9	3.0	1.4	0.2	0
2	4.7	3.2	1.3	0.2	0
3	4.6	3.1	1.5	0.2	0
4	5.0	3.6	1.4	0.2	0

Step 3: Splitting the training and testing data
```python
# Import train_test_split function

X=data[['sepal length', 'sepal width', 'petal length', 'petal
width']] # Features
y=data['species'] # Labels

# Split dataset into training set and test set
X_train, X_test, y_train, y_test = train_test_split (X,
y, test_size=0.3) # 70% training and 30% test
```

Step 4: importing model and compilation
```python
#Create a Gaussian Classifier
clf=RandomForestClassifier (n_estimators=100)

#Train the model using the training sets y_pred=clf.predict
(X_test)
```

```
clf.fit (X _ train,y _ train)

y _ pred=clf.predict (X _ test)
```

Step 5: Seeing model accuracy and prediction
```
# Model Accuracy, how often is the classifier correct?
print ("Accuracy:",metrics.accuracy _ score (y _ test, y _ pred))
```

```
Output:
Accuracy: 1.0

species _ idx = clf.predict ([[3, 5, 4, 2]])[0]
iris.target _ names[species _ idx]
```

Output:
```
'versicolor'
```

Pros:

- For classification and regression functions, the random forest classifier may be used.
- The missing values can be managed.
- Even if we have more trees in the forest, this will not over suit the pattern.

Cons:

- Inherently, an ensemble model is less interpretable than a decision tree.
- A large number of deep trees can cost a great deal (but can be paralleled) and use a great deal of memory.
- Predictions are slower, which can pose implementation problems.

Summary

Machine learning algorithms are being used for so many purposes and which help us to get useful result by analyzing the data available from several resources. Here we have the different algorithms of machine learning like linear regression, logistic regression, decision tree etc., which have discussed in this chapter with implementation using real time dataset name Irish Dataset which is most famous dataset which consist of data of flowers. From the implementation part of the algorithm we can understand how these algorithms can be implemented in real time. pros and cons of algorithms also have been stated for better understanding.

Chapter 10

Disease Classification and Detection in Plants

Introduction

In agriculture, plant diseases have always been a major problem, as they cause crop quality to decrease, resulting in yield. Plant disease has a range of effects from minor symptoms to serious damages in whole crop areas, which lead to high financial costs and significantly impact the agricultural economy, particularly in developing countries dependent on a single crop or on a few crops.

Various approaches have been developed to detect disease in order to avoid significant losses. Causal agents have been precisely defined by methods developed in molecular biology and immunology. However, for many farmers these techniques are inaccessible and require detailed knowledge of the field or substantial time and energy to be used. According to the United Nations Food and Agriculture Organization, most farms in the world are small and family-owned in developed countries. Such families provide food for a significant proportion of the population of the planet. However, hunger and food insecurity are not rare and there is restricted access to markets and services. For the above reasons, a great deal of work has been undertaken in order to establish methods that are sufficiently reliable and available to most farmers. Where Plant disease is an essential factor contributing to a major decline in plant production quality and quantity. Plant disease identification and classification are a significant role in improving plant production and economic development.

As it come to notice that how plant disease has been a major concern in the agriculture and many technologies are been used to detect it with an ease and artificial intelligence is one of them.

As an example, for plant disease detection and classification, here by using Artificial Intelligence's deep learning a model will be made and trained based on dataset from Kaggle which is plant village. It can be useful for identifying real-time plant disease In the Project we will make a CNN model which will be used for classification and detection of disease in potato based on the data we train it with.

Installing Packages

To develop a model or project on plant disease detection we will be requiring several libraries of Artificial Intelligence and python. Here we have discussed step by step their Installation and description for better understanding.

To install all the packages of libraries we will use PIP in windows. The libraries which are used to develop this project are:

- Keras:Command to install on windows: pip install keras
- NumPy: Command to install on windows: pip install keras
- Math: Command to install on windows: pip install math
- Matplotlib: Command to install on windows: pip install matplotlib
- TensorFlow: Command to install on windows: pip install tensorflow

Dataset

We are here taking a Dataset name Plant village from Kaggle which has 15 categories based on different crop diseases. We have selected to detect disease in potato plant for which we have 3 categories like Early Blight, Late Blight and healthy potato. For each these classes we will be having images as an input to train the model for prediction.

Dataset can be download from here. https://www.kaggle.com/emmarex/plantdisease

Implementation Steps

In the very first step of disease detection we will first import all the libraries required for the development for project. In this project development google research Collaboratory has been used and it can be accessed by this link http://colab.research.google.com/ and you can also use the Anaconda for running this project but please specify all the path related to dataset and testing images into the project.

Step 1: First we will import the basic libraries which are required to develop this disease detection project are

```
import numpy as np
import math
import keras
import matplotlib.pyplot as plt
from tensorflow.keras.layers import *
from tensorflow.keras.preprocessing.image import
ImageDataGenerator
from tensorflow.keras import models
from tensorflow.keras.models import Model,Sequential
from tensorflow.keras.optimizers import *
from tensorflow.keras.callbacks import LearningRateScheduler
```

Output: Using TensorFlow backend.

These are all the libraries which will be required and processed through the project for various purposes.

Step 2: There are two methods by which we can import the data for the project
First method:

```
!pip install kaggle
!mkdir .kaggle
```

```
import json
token = {"username":"","key":""}
with open ('/content/.kaggle/kaggle.json', 'w') as file:
json.dump (token, file)

!chmod 600 /content/.kaggle/kaggle.json
!cp /content/.kaggle/kaggle.json /root/.kaggle/kaggle.json
!kaggle datasets download -d xabdallahali/plantvillage-dataset/
color -p /content
```

Output:
```
Downloading plantvillage-dataset.zip to /content
99% 2.02G/2.04G [00:44<00:00, 0.6MB/s]
100% 2.04G/2.04G [00:44<00:00, 1.6MB/s]
```

Here we directly download the dataset from the Kaggle you can use it as it in your file or you can use the second method.

Second Method: If you are using the google research Collaboratory then you can download the dataset from this link https://www.kaggle.com/emmarex/plantdisease and upload it in your google drive "Colab Notebooks" folder then you can link it by using this code

```
from google.colab import drive
drive.mount ('/gdrive')
```

Output:
Go to this URL in a browser: https://accounts.google.com/o/oauth2/auth?client_id=94.................y

Enter your authorization code:

....

Mounted at /gdrive

This is the code for mounting your drive with google research Collaboratory which will ask you authentication which you can do it with your Gmail account by just following the simple steps. As you can see in the output how Collaboratory will ask you to verify it by clinking in the given link then it will get mounted by giving a authentication to it.

```
with open ('/gdrive/My Drive/foo.txt', 'w') as f:
f.write ('Hello Google Drive!')
!cat '/gdrive/My Drive/foo.txt'
```

Output:
Hello Google Drive!

This is the code which will help you to check whether your drive has been mounted with the Collaboratory or not.

After doing this we will access our dataset which has been uploaded in the drive directly and unzip it to access the files which we require in this project.

```
from zipfile import ZipFile
# specifying the zip file name
file_name = "plantvillage-dataset.zip" #path for location of
dataset
# opening the zip file in READ mode
with ZipFile (file_name, 'r') as zi:
# extracting all the files
print ('Extracting all the files now...')
zi.extractall ()
print ('Done!')
```

ouput:
```
Extracting all the files now...
Done!
```

Here we are simply extracting all the data from the ZIP folder of PlantVillage for further use.

```
#splitting data into train and test
!pip install split-folders tqdm
```

Output:
```
Collecting split-folders
```
Downloading https://files.pythonhosted.org/
packages/20/67/29dda743e6d23ac1ea3d16704d8bbb48d65faf3f1b1eaf53153b3
da56c56/split_folders-0.3.1-py3-none-any.whl

Requirement already satisfied: tqdm in /usr/local/lib/python3.6/dist-packages (4.38.0)

Installing collected packages: split-folders

Successfully installed split-folders-0.3.1

```
!mkdir potato
! cp -r '/content/plantvillage dataset/segmented/Potato_ _ _
Early_blight' 'potato/early blight'
! cp -r '/content/plantvillage dataset/segmented/Potato_ _ _
Late_blight' 'potato/late blight'
! cp -r '/content/plantvillage dataset/segmented/Potato_ _ _
healthy' 'potato/healthy'
```

Here we are making the directories for potato disease detection into the folder which is been extracted from plant village dataset.

```
import split_folders
# Split with a ratio.
# To only split into training and validation set, set a tuple
to 'ratio', i.e, ' (.8, .2)'.
#split_folders.ratio ('color', output="dataset", seed=1337,
ratio= (.8, .1, .1)) # default values
split_folders.ratio ('potato', output="dataset", seed=1337,
ratio= (.8, 0.2)) # default values
```

Output:
Copying files: 2152 files [00:00, 9225.64 files/s]

Step 3: Building CNN model
First, we will build our CNN Model for which here is the code.

```
def model_cnn ():
    model = Sequential ([
        BatchNormalization (axis=1, input_shape= (256,256,3)),

        Convolution2D (32,3,3, activation='relu'),
        BatchNormalization (axis=1),
        MaxPooling2D ( (3,3)),
        Convolution2D (64,3,3, activation='relu'),
        BatchNormalization (axis=1),
        MaxPooling2D ((3,3)),
        Flatten (),
        Dense (200, activation='relu'),
        Dropout (0.2),
        BatchNormalization (),
        Dense (3, activation='softmax')
    ])
return model
```

The model type which we are using is the basic and easy to make any CNN model which is Sequential. Here we have defined whole model into a function so that we can call it whenever it is required.

And Sequential is the easiest method to build it by using Keras. It allows us to build the CNN model Layer by layer.

In the Model, our first Layer is the Convolution2D (32,3,3, Activation='relu'), It deals with the input images and the argument 32 Depicts the number of nodes in the layer which can be changed from time to time based on dataset size. Where the 3,3 in the argument depicts the kernel size which is the size of the filter matrix for our convolution layer in the model.so a kernel size of 3 means we have 3X3 filter matrix for our model layer. The Argument Activation is the activation function for the layer which is ReLU or Rectified Liner Activation which works good for the neural networks.

After that we will use Batch Normalization (axis=1) for normalizing and re scaling the data where argument axis=1 represents the channel. After which we will use max pooling 2D for selecting maximum features covered by the filters.

In the Next layer Convolution2D (64,3,3, Activation='relu') where In this layer we will have 64 nodes for the input images to process and continuously we will have both batch normalization and max pooling again.

In between the layers we have "flatten () "layer which always server as the connection between the convolution and dense layer.

In the model we will have 2 output layers where first layer As a output layer we have Dense (200, activation='reli') layer. Dense is a standard type in neural network which can be used for many purposes. Here the argument 200 in dense layer means we will have 200 nodes in our output layer one for each possible outcome between 0-199. After this we are using dropout (0.2) which is used to drop the certain units from the CNN model which will help us to prevent over fitting in the model and an approach to regularization in the network which will help in reducing the interdependencies among the nodes of network. After this we will use batch Normalization in the layer for normalizing and re scaling the data. And the we have our final and 2nd output layer Dense (3, activation='softmax') which has 3 modes for every output from 0-2 and activation function as softmax.

We can also code the same model by using add function which is used to code layer by layer of CNN model.

Alternative code for above built model using add function:

Step 4: Now we will generate the data for model and then do the training.

```
train _ datagen = ImageDataGenerator (rescale=1./255)

validation _ datagen = ImageDataGenerator (rescale=1./255)
image _ size = 256
# Change the batchsize according to your system RAM
train _ batchsize = 20
val _ batchsize = 20
train _ generator = train _ datagen.flow _ from _ directory (
    'dataset/train',
    target _ size= (image _ size, image _ size),
    batch _ size=train _ batchsize,
    class _ mode='categorical')

validation _ generator = validation _ datagen.flow _ from _
                                            directory (
    'dataset/val',
    target _ size= (image _ size, image _ size),
    batch _ size=val _ batchsize,
    class _ mode='categorical',
    shuffle=True)
```

Output:

Found 1721 images belonging to 3 classes.

Found 431 images belonging to 3 classes

Here we are generating the dataset related to potato and dividing them into training and testing sets for our CNN model. As you can see in the code, we have declared image size and batch size which can altered according to the need of a dataset and system.

Here we are making a function which has learning rate which will get change based on epoch changes so that we can ger better result from the trained model.

Note: We can also keep a constant learning rate for all the epochs. It is totally our wish to do so. Sometimes if we don't get required accuracy while testing the model we do changes in the learning rate.

```
def step _ decay (epoch):
if epoch<=6:
return 1e-3
elif epoch<=20:
return 1e-4
else:
return 1e-5


lrate = LearningRateScheduler (step _ decay)
```

Here will recall our model and start training the model by setting learning rate and optimizers values for better results which you can see in the outputs.

```
#define model
model = model _ cnn ()


#calling the learning rate
callbacks = [lrate]


# Compile the model
model.compile (loss='categorical _ crossentropy',
optimizer=Adam (lr=0.0001, beta _ 1=0.9, beta _ 2=0.999,
    epsilon=1e-08, decay=0.0), metrics=['acc'])
history = model.fit _ generator (
    train _ generator,
    steps _ per _ epoch=train _ generator.samples/train _ generator.
        batch _ size ,
    validation _ data=validation _ generator,
    validation _ steps=validation _ generator.samples/validation _
        generator.batch _ size,
    epochs=23,
    callbacks = callbacks)
```

Output:

```
Epoch 1/23
87/86 [==================] - 5s 54ms/step - loss: 0.5352 - acc:
0.7966 - val _ loss: 0.7801 - val _ acc: 0.6288 - lr: 0.0010
Epoch 2/23
87/86 [==================] - 5s 52ms/step - loss: 0.3022 - acc:
0.8879 - val _ loss: 0.6320 - val _ acc: 0.6914 - lr: 0.0010
Epoch 3/23
87/86 [==================] - 5s 53ms/step - loss: 0.2282 - acc:
0.9169 - val _ loss: 0.3148 - val _ acc: 0.9002 - lr: 0.0010
Epoch 4/23
87/86 [==================] - 5s 53ms/step - loss: 0.2011 - acc:
0.9303 - val _ loss: 0.2739 - val _ acc: 0.8585 - lr: 0.0010
Epoch 5/23
87/86 [==================] - 5s 52ms/step - loss: 0.1417 - acc:
0.9553 - val _ loss: 0.6454 - val _ acc: 0.7262 - lr: 0.0010
Epoch 6/23
87/86 [==================] - 5s 52ms/step - loss: 0.1223 - acc:
0.9599 - val _ loss: 0.1182 - val _ acc: 0.9397 - lr: 0.0010
Epoch 7/23
87/86 [==================] - 5s 53ms/step - loss: 0.0917 - acc:
0.9692 - val _ loss: 0.2812 - val _ acc: 0.8794 - lr: 0.0010
Epoch 8/23
87/86 [==================] - 5s 53ms/step - loss: 0.0757 - acc:
0.9773 - val _ loss: 0.0892 - val _ acc: 0.9722 - lr: 1.0000e-04
Epoch 9/23
87/86 [==================] - 5s 53ms/step - loss: 0.0691 - acc:
0.9791 - val _ loss: 0.0852 - val _ acc: 0.9606 - lr: 1.0000e-04
Epoch 10/23
87/86 [==================] - 5s 53ms/step - loss: 0.0526 - acc:
0.9878 - val _ loss: 0.0890 - val _ acc: 0.9722 - lr: 1.0000e-04
Epoch 11/23
87/86 [==================] - 5s 52ms/step - loss: 0.0864 - acc:
0.9843 - val _ loss: 0.0945 - val _ acc: 0.9675 - lr: 1.0000e-04
Epoch 12/23
87/86 [==================] - 5s 52ms/step - loss: 0.0527 - acc:
0.9872 - val _ loss: 0.0767 - val _ acc: 0.9698 - lr: 1.0000e-04
Epoch 13/23
87/86 [==================] - 4s 51ms/step - loss: 0.0470 - acc:
0.9872 - val _ loss: 0.1524 - val _ acc: 0.9420 - lr: 1.0000e-04
Epoch 14/23
87/86 [==================] - 4s 52ms/step - loss: 0.0486 - acc:
0.9901 - val _ loss: 0.0918 - val _ acc: 0.9652 - lr: 1.0000e-04
Epoch 15/23
87/86 [==================] - 5s 52ms/step - loss: 0.0417 - acc:
0.9895 - val _ loss: 0.0852 - val _ acc: 0.9652 - lr: 1.0000e-04
Epoch 16/23
87/86 [==================] - 4s 51ms/step - loss: 0.0420 - acc:
0.9942 - val _ loss: 0.0744 - val _ acc: 0.9768 - lr: 1.0000e-04
```

```
Epoch 17/23
87/86 [===================] - 4s 51ms/step - loss: 0.0424 - acc:
0.9919 - val _ loss: 0.0953 - val _ acc: 0.9675 - lr: 1.0000e-04
Epoch 18/23
87/86 [===================] - 4s 52ms/step - loss: 0.0372 - acc:
0.9895 - val _ loss: 0.1163 - val _ acc: 0.9606 - lr: 1.0000e-04
Epoch 19/23
87/86 [===================] - 5s 53ms/step - loss: 0.0315 - acc:
0.9965 - val _ loss: 0.0718 - val _ acc: 0.9675 - lr: 1.0000e-04
Epoch 20/23
87/86 [===================] - 4s 51ms/step - loss: 0.0371 - acc:
0.9919 - val _ loss: 0.1132 - val _ acc: 0.9698 - lr: 1.0000e-04
Epoch 21/23
87/86 [===================] - 4s 51ms/step - loss: 0.0275 - acc:
0.9948 - val _ loss: 0.0712 - val _ acc: 0.9722 - lr: 1.0000e-04
Epoch 22/23
87/86 [===================] - 4s 51ms/step - loss: 0.0345 - acc:
0.9930 - val _ loss: 0.0813 - val _ acc: 0.9768 - lr: 1.0000e-05
Epoch 23/23
87/86 [===================] - 4s 51ms/step - loss: 0.0306 - acc:
0.9942 - val _ loss: 0.0875 - val _ acc: 0.9768 - lr: 1.0000e-05
```

As you see that we have trained our model which gave us an accuracy of 99.42% and validation accuracy of 97.68%. After this we will plot our results using matplotlib.

Step 5: Summarizing the Model after training

```
# summarize history for accuracy
plt.plot (history.history['acc'])
plt.plot (history.history['val _ acc'])
plt.title ('model accuracy')
plt.ylabel ('accuracy')
plt.xlabel ('epoch')
plt.legend (['train', 'test'], loc='upper left')
plt.show ()
```

Output: as in Fig. 10.1

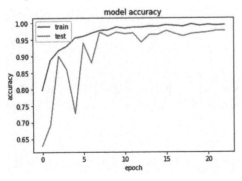

Fig. 10.1 Accuracy Representation of model

```
# summarize history for loss
plt.plot (history.history['loss'])
plt.plot (history.history['val_loss'])
plt.title ('model loss')
plt.ylabel ('loss')
plt.xlabel ('epoch')
plt.legend (['train', 'test'], loc='upper left')
plt.show ()
```

Output: as in Fig. 10.2

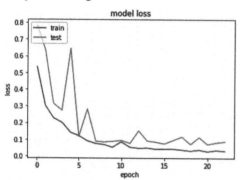

Fig. 10.2 Loss Representation of model

Here we have plotted our results between model accuracy and model loss. Which you can see in the graphs.

Step 6: Saving the model
```
model_json = model.to_json ()
with open ("model.json", "w") as json_file:
json_file.write (model_json)
# serialize weights to HDF5
model.save_weights ("model.h5")
print ("Model Saved")
```

output:
Model Saved

In this step we are saving the trained model which can be utilized further. By doing this we no need to train our model again and again for any new data input to test their results.

Step 7: Testing the model
Here we will be testing the model which we have saved in earlier step-5.

```
#importing the model
```

```
from tensorflow.keras.models import model_from_json
json_file = open ('model.json', 'r')
loaded_model_json = json_file.read ()
json_file.close ()
loaded_model = model_from_json (loaded_model_json)
# load weights into new model
loaded_model.load_weights ('model.h5')
```

First we are loading our trained model which is ready to use for plant disease detection in potato and its healthy status.

```
class_dict = train_generator.class_indices
class_list = list (class_dict.keys ())
class_list
```

here we are creating a list for disease status of plant related to its categories

```
class_list=['early blight', 'healthy', 'late blight']
```

Here we are defining the list with disease names to predict while providing the input.

```
import cv2
def preprocess (image):
image = cv2.cvtColor (image, cv2.COLOR_BGR2RGB)
image = cv2.resize (image ,  (256,256))
# create a simple mask image th the original one
mask = np.zeros (image.shape[:2], np.uint8)

# specify the background and foreground model
backgroundModel = np.zeros ( (1, 65), np.float64)
foregroundModel = np.zeros ( (1, 65), np.float64)

# define the Region of Interest  (ROI) as the coordinates of the
rectangle
rectangle =  (20, 20, 220, 220)
cv2.grabCut (image, mask,rectangle,
backgroundModel, foregroundModel,
1,cv2.GC_INIT_WITH_RECT)

mask2 = np.where ( (mask == 2)| (mask == 0), 0, 1).astype ('uint8')
image = image * mask2[:, :, np.newaxis]
image = image/255
return image
```

Here we have used Computer Vision for providing input to the trained model.

```
def predict_class (model,image_path):
```

```
original _ image = cv2.imread (image _ path)
image = preprocess (original _ image)
tab = []
tab.append (image)
tab = np.array (tab)
h = np.expand _ dims (tab, axis=2)
h=model.predict (tab)
predicted _ class = np.argmax (h[0])
confidence = np.max (h[0])

# plot resutls
plt.figure (figsize =  (4,4))
plt.imshow (image)
plt.axis ('off')
plt.title ('prediction:  '+class _ list[predicted _ class]+'  with
conf '+str (confidence))
plt.show ()
```

here we are plotting the results for the input data images into the model.
```
predict _ class   (loaded _ model,'test/lateb.jpg')#image   path   is
provided for prediction
predict _ class (loaded _ model,'test/lateb2.jpg')
```

Output:
prediction late blight with conf 0.9999802

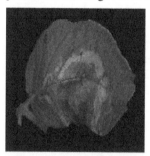

prediction late blight with conf 0.99996245

predict_class (loaded_model,'test/earlyb.jpg')
predict_class (loaded_model,'test/earlyb1.jpg')

Output:
prediction late blight with conf 0.9517815

prediction late blight with conf 9.5648595

predict_class (loaded_model,'test/healthy.jpg')
predict_class (loaded_model,'test/healthy1.jpg')

Output:
prediction late blight with conf 0.5035014

prediction late blight with conf 0.5035014

As you have in the project the how a CNN model is mad for plant disease detection and how we have saved the model for further future use. You can download the saved model from this link and can directly use it for prediction. # link is_____.

Summary

Plant disease classification has been a very prominent in agriculture which make it more useful for farmer and identify and to prevent it being spread among the plant so that damage to the crops can be prevented as in this chapter we have used plant village dataset to identify the disease in plant using the symptoms of it on plant leaves using machine learning, CNN and computer vision which are the domains of artificial Intelligence. The current project which have been made in this chapter can be used in real time for identifying diseases in potato plants. In the same way we can design an recognition and identification of diseases in plant can be designed for every plant which can benefit farmers in their farming process.

Chapter 11

Species Recognition in Flowers

Introduction

The conventional approach of the human person for plant classification is comparing the color and shape of the leaves, but by using Artificial Intelligence analysis can provide quicker and more detailed results by analyzing the morphology of the leaves and provides more details about the leaf properties. Species Recognition has been a very important part of agriculture and there are thousands of species in plants which can be identified by a single person and even if it does so It will take a huge lot of time to do it.

In this chapter we will discuss and built a project related to identify species in flower with an image. In this project we will be using Artificial Intelligence different branches like machine learning, deep learning and computer vision to build such an amazing project. So, let's begin

Installation of Packages

In this species Recognition we have used so many libraries which need be installed before implementing this project. The command for installing these libraries are

- Warnings- pip install pytest-warnings/warnings
- NumPy- pip install Numpy
- hPandas- pip install pandas
- Matplotlib- pip install matplotlib
- Seaborn- pip install seaborn
- Tqdm- pip install tqdm
- Pillow/PIL- pip install pillow
- TensorFlow- pip install TensorFlow
- Keras- pip Install keras

For Python IDE: Enter these provided commands on the command prompt in windows
For Anaconda: Enter these commands on Anaconda Prompt

If you are using any other Environment or IDE then you have to specify the path for python and you can do the installation.

This project was implemented in Google Research Collaboratory. And you can download the copy of the code form this link_____

Dataset

In this project we are using a dataset from Kaggle which has 5 categories of species in flowers namely daisy, dandelion, rose, sunflower and tulip which totally consist of 4242 images with 320×240 pixels. The images in the dataset is based on the data flicr, google images and Yandex images. These images can be used to recognize the plant related to these species.

You can download the same dataset from this link _____

Implementation:

Note: This project is purely implemented on google research Collaboratory which you access by this link_____ and you are free to use any other IDE for project implementation.

Step 1: Importing all the Libraries

Note: The Symbol # represents the comment in the project which will make you understand about the code.

```
# Ignore the warnings
import warnings
warnings.filterwarnings ('always')
warnings.filterwarnings ('ignore')


# data visualisation and manipulation
import numpy as np
import pandas as pd
import matplotlib.pyplot as plt
from matplotlib import style
import seaborn as sns


#configure
# sets matplotlib to inline and displays graphs below the
corressponding cell.
% matplotlib inline
style.use ('fivethirtyeight')
sns.set (style='whitegrid',color _ codes=True)


#model selection
from sklearn.model _ selection import train _ test _ split
from sklearn.model _ selection import KFold
from sklearn.metrics import accuracy _ score,precision _
score,recall _ score,confusion _ matrix,roc _ curve,
    roc _ auc _ score
from sklearn.model _ selection import GridSearchCV
from sklearn.preprocessing import LabelEncoder
```

```
#preprocess.
from keras.preprocessing.image import ImageDataGenerator

#These are deep learning libraries which is present in keras
if you have installed keras these all will get installed
automatically.
from keras import backend as K
from keras.models import Sequential
from keras.layers import Dense
from keras.optimizers import Adam,SGD,Adagrad,Adadelta,RMSprop
from keras.utils import to_ categorical

# we use these libraries specifically for CNN (convolutional
Neural Network) Building which will be used to use to recognize
the species of flower.
from keras.layers import Dropout, Flatten,Activation
from keras.layers import Conv2D, MaxPooling2D,
BatchNormalization

import tensorflow as tf
import random as rn

# here we are using OpenCV library to give input to the trained
model and specifically for manipulating zipped images and
getting NumPy arrays of pixel values of images.
import cv2
import numpy as np
from tqdm import tqdm
import os
from random import shuffle
from zipfile import ZipFile
from PIL import Image
```

Step 2: Mounting to drive for dataset

If you are using the google research Collaboratory then you can download the dataset from this link _____ and upload it in your google drive "Colab Notebooks" folder then you can link it by using this code

```
from google.colab import drive
drive.mount ('/gdrive')
```

Output:

Go to this URL in a browser: https://accounts.google.com/o/oauth2/auth?client_ id=94..................y

Enter your authorization code:

....

Mounted at /gdrive

This is the code for mounting your drive with google research Collaboratory which will ask you authentication which you can do it with your Gmail account by just following the simple steps. As you can see in the output how Collaboratory will ask you to verify it by clinking in the given link then it will get mounted by giving a authentication to it.

```
with open ('/gdrive/My Drive/foo.txt', 'w') as f:
f.write ('Hello Google Drive!')
!cat '/gdrive/My Drive/foo.txt'
```

Output:
Hello Google Drive!

This is the code which will help you to check whether your drive has been mounted with the Collaboratory or not.

After doing this we will access our dataset which has been uploaded in the drive directly and unzip it to access the files which we require in this project.

Step 3: Getting Data ready for Model
Here we are forming an array of images X and Y which will store all the images in the form of training and validation images for the CNN model to train and get tested with real data.

```
X=[]
Z=[]
IMG _ SIZE=150 #assigning the image size constantly for the project
FLOWER _ DAISY _ DIR='drive/Colab Notebooks/flower recog/flowers/
daisy'
FLOWER _ SUNFLOWER _ DIR='drive/Colab     Notebooks/flower     recog/
flowers/sunflower'
FLOWER _ TULIP _ DIR='drive/Colab Notebooks/flower recog/flowers/
tulip'
FLOWER _ DANDI _ DIR='drive/Colab Notebooks/flower recog/flowers/
dandelion'
FLOWER _ ROSE _ DIR='drive/Colab     Notebooks/flower     recog/flowers/
rose'
```

Creating a function to label the images in the data which will be further used for training purpose.

```
def assign _ label (img,flower _ type):
return flower _ type
```

here after labelling we will make a training data Separately from each categories of flower

```
def make _ train _ data (flower _ type,DIR):
    for img in tqdm (os.listdir (DIR)):
    label=assign _ label (img,flower _ type)
```

```
        path = os.path.join (DIR,img)
        img = cv2.imread (path,cv2.IMREAD _ COLOR)
        img = cv2.resize (img, (IMG _ SIZE,IMG _ SIZE))

        X.append (np.array (img))
        Z.append (str (label))

# Training data for daisy
make _ train _ data ('Daisy',FLOWER _ DAISY _ DIR)
print (len (X))
```

Output:

100%|████████████| 769/769 [00:07<00:00, 101.93it/s]769

```
#Training data for sunflower
make _ train _ data ('Sunflower',FLOWER _ SUNFLOWER _ DIR)
print (len (X))
```

Output:

100%|████████████| 766/766 [00:07<00:00, 176.83it/s]1535

```
#training Data for Tulip
make _ train _ data ('Tulip',FLOWER _ TULIP _ DIR)
print (len (X))
```

Output:

100%|████████████| 991/991 [00:08<00:00, 116.01it/s]2526

```
#trining data for dandelion
make _ train _ data ('Dandelion',FLOWER _ DANDI _ DIR)
print (len (X))
```

Output:

100%|████████████| 1052/1052 [00:09<00:00, 114.67it/s]3578

```
#training data for Rose
make _ train _ data ('Rose',FLOWER _ ROSE _ DIR)
print (len (X))
```

Output:

100%|████████████| 784/784 [00:06<00:00, 116.67it/s]4362

This is how training data for all the flowers is ready for the model to get trained.

Step 4: Visualizing the data

After Preparing the data for training we will visualize the data we have prepared using matplotlib and opencv.

```
fig,ax=plt.subplots (5,2)
fig.set _ size _ inches (15,15)
for i in range (5):
    for j in range (2):
      l=rn.randint (0,len (Z))
      ax[i,j].imshow (X[l])
      ax[i,j].set _ title ('Flower: '+Z[l])
plt.tight _ layout ()
```

Output: as I Fig. 11.1

Fig. 11.1 Data visualization from dataset

Step-5: Labelling the data and splitting it into training and validation data

```
#labelling
le=LabelEncoder ()
Y=le.fit _ transform (Z)
Y=to _ categorical (Y,5)
X=np.array (X)
X=X/255

#After labelling we will do Splitting of data into Training and
Validation Sets
x _ train,x _ test,y _ train,y _ test=train _ test _ split  (X,Y,test _
size=0.25,random _ state=42)

#Setting the random seeds
np.random.seed (42)
rn.seed (42)
tf.set _ random _ seed (42)
```

Step-6: Building The model

For building the model I have used Keras Sequential in which it is very easy to build any model for which we have a lot of images as input to extract features and classify them.

In the model creation, I have created four convolutional layer, In the first layer we have 32 filters, 5×5 of kernel matrix with padding and an activation function ReLU with an input size of an image.

In the Second layer we have 64, filters, 3×3 of kernel matrix with padding and an activation function ReLU with an input size of an image.

In the Third and fourth layer we have 96, filters, 3×3 of kernel matrix with padding and an activation function ReLU with an input size of an image.

Following all these convolutional layers we will have a max pooling layer which will help us to get maximum features from the filters. In between the layers we have "flatten () "layer which always server as the connection between the convolution and dense layer. And at last we have our fully connected output layers i.e.,, dense layers with an activation function.

```
# modelling starts using a CNN.

model = Sequential ()
model.add (Conv2D (filters = 32, kernel _ size = (5,5),padding =
'Same',activation ='relu', input _ shape = (150,150,3)))
model.add (MaxPooling2D (pool _ size= (2,2)))


model.add (Conv2D (filters = 64, kernel _ size = (3,3),padding =
'Same',activation ='relu'))
model.add (MaxPooling2D (pool _ size= (2,2), strides= (2,2)))
```

```
model.add (Conv2D (filters =96, kernel _ size = (3,3),padding =
'Same',activation ='relu'))
model.add (MaxPooling2D (pool _ size= (2,2), strides= (2,2)))

model.add (Conv2D (filters = 96, kernel _ size = (3,3),padding =
'Same',activation ='relu'))
model.add (MaxPooling2D (pool _ size= (2,2), strides= (2,2)))

model.add (Flatten ())

model.add (Dense (512))
model.add (Activation ('relu'))
model.add (Dense (5, activation = "softmax"))
```

#after the model building we will declare our number of epochs and their batch size
and we are using LR annealer for fixing our learning rate after a certain period of
time while compiling the model.

```
#using a LR Annealer
batch _ size=128
epochs=50

from keras.callbacks import ReduceLROnPlateau
red _ lr= ReduceLROnPlateau (monitor='val _ acc',
patience=3,verbose=1,factor=0.1)
```

#we are here performing Data Augmentation to prevent Overfitting

datagen = ImageDataGenerator (

featurewise_center=False, # set input mean to 0 over the dataset

samplewise_center=False, # set each sample mean to 0

featurewise_std_normalization=False, # divide inputs by std of the dataset

samplewise_std_normalization=False, # divide each input by its std

zca_whitening=False, # apply ZCA whitening

rotation_range=10, # randomly rotate images in the range (degrees, 0 to 180)

zoom_range = 0.1, # Randomly zoom image

width_shift_range=0.2, # randomly shift images horizontally (fraction of total width)

height_shift_range=0.2, # randomly shift images vertically (fraction of total height)

horizontal_flip=True, # randomly flip images

vertical_flip=False) # randomly flip images

datagen.fit (x_train)

#now we will do Compiling of our Model & Viewing Summary of Model

```
model.compile (optimizer=Adam (lr=0.001),loss='categorical_
crossentropy',metrics=['accuracy'])
model.summary ()
```

Output:

```
_ _ _ _ _ _ _ _ _ _ _ _ _ _ _ _ _ _ _ _ _ _ _ _ _ _ _ _ _
Layer (type) Output Shape Param #
=================================================================
conv2d_1 (Conv2D) (None, 150, 150, 32) 2432
_ _ _ _ _ _ _ _ _ _ _ _ _ _ _ _ _ _ _ _ _ _ _ _ _ _ _ _ _
max_pooling2d_1 (MaxPooling2 (None, 75, 75, 32) 0
_ _ _ _ _ _ _ _ _ _ _ _ _ _ _ _ _ _ _ _ _ _ _ _ _ _ _ _ _
conv2d_2 (Conv2D) (None, 75, 75, 64) 18496
_ _ _ _ _ _ _ _ _ _ _ _ _ _ _ _ _ _ _ _ _ _ _ _ _ _ _ _ _
max_pooling2d_2 (MaxPooling2 (None, 37, 37, 64) 0
_ _ _ _ _ _ _ _ _ _ _ _ _ _ _ _ _ _ _ _ _ _ _ _ _ _ _ _ _
conv2d_3 (Conv2D) (None, 37, 37, 96) 55392
_ _ _ _ _ _ _ _ _ _ _ _ _ _ _ _ _ _ _ _ _ _ _ _ _ _ _ _ _
max_pooling2d_3 (MaxPooling2 (None, 18, 18, 96) 0
_ _ _ _ _ _ _ _ _ _ _ _ _ _ _ _ _ _ _ _ _ _ _ _ _ _ _ _ _
conv2d_4 (Conv2D) (None, 18, 18, 96) 83040
_ _ _ _ _ _ _ _ _ _ _ _ _ _ _ _ _ _ _ _ _ _ _ _ _ _ _ _ _
max_pooling2d_4 (MaxPooling2 (None, 9, 9, 96) 0
_ _ _ _ _ _ _ _ _ _ _ _ _ _ _ _ _ _ _ _ _ _ _ _ _ _ _ _ _
flatten_1 (Flatten) (None, 7776) 0
_ _ _ _ _ _ _ _ _ _ _ _ _ _ _ _ _ _ _ _ _ _ _ _ _ _ _ _ _
dense_1 (Dense) (None, 512) 3981824
_ _ _ _ _ _ _ _ _ _ _ _ _ _ _ _ _ _ _ _ _ _ _ _ _ _ _ _ _
activation_1 (Activation) (None, 512) 0
_ _ _ _ _ _ _ _ _ _ _ _ _ _ _ _ _ _ _ _ _ _ _ _ _ _ _ _ _
dense_2 (Dense) (None, 5) 2565
=================================================================
Total params: 4,143,749
Trainable params: 4,143,749
Non-trainable params: 0
```

_ _

Step 7: Training the Model with dataset
After the compilation of the model we will train our model with our testing data which we have labella and separated earlier.

```
History = model.fit_generator (datagen.flow (x_train,y_train,
batch_size=batch_size),
epochs = epochs, validation_data = (x_test,y_test),
verbose = 1, steps_per_epoch=x_train.shape[0] // batch_size)
# model.fit (x_train,y_train,epochs=epochs,batch_
size=batch_size,validation_data = (x_test,y_test))
```

Output:
```
Epoch 1/50
25/25 [==============================] - 23s 903ms/step - loss:
1.4743 - acc: 0.3588 - val_loss: 1.2462 - val_acc: 0.5069
Epoch 2/50
25/25 [==============================] - 20s 808ms/step - loss:
1.1573 - acc: 0.5212 - val_loss: 1.1553 - val_acc: 0.5252
Epoch 3/50
25/25 [==============================] - 20s 812ms/step - loss:
1.0605 - acc: 0.5745 - val_loss: 1.0670 - val_acc: 0.6013
Epoch 4/50
25/25 [==============================] - 21s 820ms/step - loss:
1.0024 - acc: 0.5956 - val_loss: 0.9759 - val_acc: 0.6114
Epoch 5/50
25/25 [==============================] - 20s 795ms/step - loss:
0.9267 - acc: 0.6437 - val_loss: 0.9823 - val_acc: 0.6205
Epoch 6/50
25/25 [==============================] - 20s 784ms/step - loss:
0.9096 - acc: 0.6477 - val_loss: 0.9942 - val_acc: 0.6004
Epoch 7/50
25/25 [==============================] - 20s 807ms/step - loss:
0.8594 - acc: 0.6677 - val_loss: 0.9092 - val_acc: 0.6581
Epoch 8/50
25/25 [==============================] - 20s 812ms/step - loss:
0.8468 - acc: 0.6779 - val_loss: 0.8853 - val_acc: 0.6783
Epoch 9/50
25/25 [==============================] - 20s 810ms/step - loss:
0.7789 - acc: 0.7019 - val_loss: 0.8087 - val_acc: 0.6856
Epoch 10/50
25/25 [==============================] - 20s 798ms/step - loss:
0.7613 - acc: 0.7089 - val_loss: 0.8019 - val_acc: 0.6939
Epoch 11/50
25/25 [==============================] - 20s 795ms/step - loss:
0.7660 - acc: 0.6997 - val_loss: 0.8029 - val_acc: 0.7030
Epoch 12/50
25/25 [==============================] - 20s 794ms/step - loss:
0.7742 - acc: 0.7022 - val_loss: 0.9206 - val_acc: 0.6480
Epoch 13/50
25/25 [==============================] - 20s 794ms/step - loss:
0.7367 - acc: 0.7154 - val_loss: 0.7945 - val_acc: 0.6929
Epoch 14/50
25/25 [==============================] - 19s 778ms/step - loss:
0.7020 - acc: 0.7354 - val_loss: 0.7705 - val_acc: 0.6948
Epoch 15/50
25/25 [==============================] - 20s 807ms/step - loss:
0.6742 - acc: 0.7325 - val_loss: 0.7577 - val_acc: 0.7140
Epoch 16/50
```

```
25/25 [==============================] - 20s 781ms/step - loss:
0.6758 - acc: 0.7418 - val _ loss: 0.7766 - val _ acc: 0.7131
Epoch 17/50
25/25 [==============================] - 20s 810ms/step - loss:
0.6605 - acc: 0.7466 - val _ loss: 0.7200 - val _ acc: 0.7269
Epoch 18/50
25/25 [==============================] - 20s 793ms/step - loss:
0.6453 - acc: 0.7500 - val _ loss: 0.7373 - val _ acc: 0.7186
Epoch 19/50
25/25 [==============================] - 20s 808ms/step - loss:
0.6381 - acc: 0.7585 - val _ loss: 0.7292 - val _ acc: 0.7269
Epoch 20/50
25/25 [==============================] - 20s 795ms/step - loss:
0.6110 - acc: 0.7673 - val _ loss: 0.7906 - val _ acc: 0.7131
Epoch 21/50
25/25 [==============================] - 20s 795ms/step - loss:
0.6187 - acc: 0.7629 - val _ loss: 0.7009 - val _ acc: 0.7296
Epoch 22/50
25/25 [==============================] - 20s 797ms/step - loss:
0.5574 - acc: 0.7865 - val _ loss: 0.6847 - val _ acc: 0.7599
Epoch 23/50
25/25 [==============================] - 20s 799ms/step - loss:
0.5884 - acc: 0.7788 - val _ loss: 0.6892 - val _ acc: 0.7415
Epoch 24/50
25/25 [==============================] - 20s 796ms/step - loss:
0.5553 - acc: 0.7869 - val _ loss: 0.7059 - val _ acc: 0.7360
Epoch 25/50
25/25 [==============================] - 20s 797ms/step - loss:
0.5731 - acc: 0.7867 - val _ loss: 0.6628 - val _ acc: 0.7479
Epoch 26/50
25/25 [==============================] - 20s 799ms/step - loss:
0.5141 - acc: 0.7989 - val _ loss: 0.6582 - val _ acc: 0.7681
Epoch 27/50
25/25 [==============================] - 20s 800ms/step - loss:
0.4982 - acc: 0.8122 - val _ loss: 0.8243 - val _ acc: 0.7232
Epoch 28/50
25/25 [==============================] - 20s 801ms/step - loss:
0.5484 - acc: 0.7898 - val _ loss: 0.6767 - val _ acc: 0.7562
Epoch 29/50
25/25 [==============================] - 20s 800ms/step - loss:
0.4928 - acc: 0.8129 - val _ loss: 0.7491 - val _ acc: 0.7498
Epoch 30/50
25/25 [==============================] - 20s 799ms/step - loss:
0.4768 - acc: 0.8238 - val _ loss: 0.6758 - val _ acc: 0.7544
Epoch 31/50
25/25 [==============================] - 20s 795ms/step - loss:
0.4678 - acc: 0.8219 - val _ loss: 0.7470 - val _ acc: 0.7415
Epoch 32/50
```

```
25/25 [==============================] - 20s 798ms/step - loss:
0.4604 - acc: 0.8260 - val_loss: 0.6842 - val_acc: 0.7489
Epoch 33/50
25/25 [==============================] - 20s 796ms/step - loss:
0.4755 - acc: 0.8263 - val_loss: 0.7710 - val_acc: 0.7589
Epoch 34/50
25/25 [==============================] - 20s 800ms/step - loss:
0.4409 - acc: 0.8345 - val_loss: 0.6872 - val_acc: 0.7782
Epoch 35/50
25/25 [==============================] - 20s 807ms/step - loss:
0.4372 - acc: 0.8378 - val_loss: 0.6577 - val_acc: 0.7626
Epoch 36/50
25/25 [==============================] - 20s 810ms/step - loss:
0.3909 - acc: 0.8534 - val_loss: 0.6526 - val_acc: 0.7910
Epoch 37/50
25/25 [==============================] - 20s 784ms/step - loss:
0.3928 - acc: 0.8558 - val_loss: 0.7110 - val_acc: 0.7663
Epoch 38/50
25/25 [==============================] - 20s 793ms/step - loss:
0.3955 - acc: 0.8530 - val_loss: 0.7136 - val_acc: 0.7773
Epoch 39/50
25/25 [==============================] - 20s 808ms/step - loss:
0.3658 - acc: 0.8659 - val_loss: 0.6593 - val_acc: 0.7773
Epoch 40/50
25/25 [==============================] - 20s 798ms/step - loss:
0.3916 - acc: 0.8588 - val_loss: 0.6378 - val_acc: 0.7782
Epoch 41/50
25/25 [==============================] - 20s 797ms/step - loss:
0.3722 - acc: 0.8565 - val_loss: 0.7199 - val_acc: 0.7736
Epoch 42/50
25/25 [==============================] - 20s 782ms/step - loss:
0.3746 - acc: 0.8533 - val_loss: 0.6931 - val_acc: 0.7791
Epoch 43/50
25/25 [==============================] - 20s 795ms/step - loss:
0.3255 - acc: 0.8787 - val_loss: 0.6562 - val_acc: 0.8029
Epoch 44/50
25/25 [==============================] - 20s 809ms/step - loss:
0.3306 - acc: 0.8738 - val_loss: 0.7102 - val_acc: 0.7635
Epoch 45/50
25/25 [==============================] - 20s 780ms/step - loss:
0.2981 - acc: 0.8809 - val_loss: 0.7033 - val_acc: 0.7929
Epoch 46/50
25/25 [==============================] - 20s 809ms/step - loss:
0.3009 - acc: 0.8903 - val_loss: 0.6488 - val_acc: 0.8066
Epoch 47/50
25/25 [==============================] - 20s 783ms/step - loss:
0.3078 - acc: 0.8829 - val_loss: 0.6670 - val_acc: 0.7965
Epoch 48/50
```

```
25/25 [==============================] - 20s 809ms/step - loss:
0.2886 - acc: 0.8906 - val_loss: 0.7584 - val_acc: 0.7828
Epoch 49/50
25/25 [==============================] - 20s 797ms/step - loss:
0.3103 - acc: 0.8860 - val_loss: 0.6943 - val_acc: 0.7809
Epoch 50/50
25/25 [==============================] - 20s 808ms/step - loss:
0.3007 - acc: 0.8898 - val_loss: 0.7433 - val_acc: 0.7754
```

After the compilation we can see the result of our model which is accuracy is 88.98% and Val. Acc is 77.54% which can be increased if provide more data for training or do some changes with the model by changing its optimizers or activation function or learning rate.

Step 8: Viewing the model performance
Now we will plot our trained model results in a graph with the help of matplotlib.

```
plt.plot (History.history['loss'])
plt.plot (History.history['val_loss'])
plt.title ('Model Loss')
plt.ylabel ('Loss')
plt.xlabel ('Epochs')
plt.legend (['train', 'test'])
plt.show ()
```

Output: as in Fig. 11.2

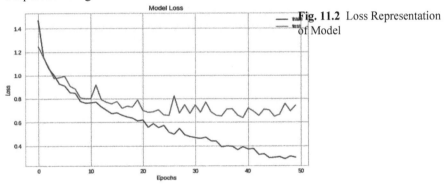

Fig. 11.2 Loss Representation of Model

```
plt.plot (History.history['acc'])
plt.plot (History.history['val_acc'])
plt.title ('Model Accuracy')
plt.ylabel ('Accuracy')
plt.xlabel ('Epochs')
plt.legend (['train', 'test'])
plt.show ()
```

Output: as in Fig. 11.3

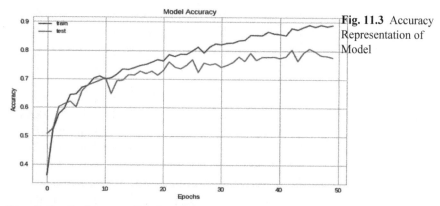

Fig. 11.3 Accuracy Representation of Model

Step 9: Prediction on validation data

Here we will get our trained model to predict the images with the validation data which we have separated while labelling and splitting our data.

```
# getting predictions on val set.
pred=model.predict (x _ test)
pred _ digits=np.argmax (pred,axis=1)

# now storing some properly as well as misclassified indexes.
i=0
prop _ class=[]
mis _ class=[]
```

Here we are importing images from the validation data to test the model for its prediction.

```
for i in range (len (y _ test)):
    if (np.argmax (y _ test[i])==pred _ digits[i]):
      prop _ class.append (i)
    if (len (prop _ class)==8):
      break

i=0
for i in range (len (y _ test)):
    if (not np.argmax (y _ test[i])==pred _ digits[i]):
      mis _ class.append (i)
    if (len (mis _ class)==8):
      break

warnings.filterwarnings ('always')
warnings.filterwarnings ('ignore')
count=0
fig,ax=plt.subplots (4,2)
fig.set _ size _ inches (15,15)
for i in range (3):
    for j in range (2):
```

```
    ax[i,j].imshow (x _ test[prop _ class[count]])
    ax[i,j].set _ title ("Predicted Flower : "+str (le.inverse _
transform ([pred_ digits[prop _ class[count]]]))+"\n"+"Actual
Flower : "+str (le.inverse _ transform (np.argmax ([y _
test[prop _ class[count]]]))))
    plt.tight _ layout ()
    count+=1
```

Output: as in Fig. 11.4

Fig. 11.4 Prediction of flowers

Step 10: Saving the model

This is the way to save a model which we have trained

```
model _ json = model.to _ json ()
with open ("model.json", "w") as json _ file:
json _ file.write (model _ json)
# serialize weights to HDF5
model.save _ weights ("model.h5")
print ("Model Saved")
```

After saving this model we can predict the values under these categories just by using the images of those species.

Summary

Species Recognition also plays an important role in the field of agriculture which can help a lot of people to identify the plant species digitally and it will last long with a guarantee that our system is correct to identify a species of plant. In this chapter we have worked on one more project which is related to species recognition and we can totally understand how to use the dataset with machine learning, deep learning and computer vision for better understand and real time use of artificial Intelligence. In the Species recognition we have used flower dataset to train our model for classification and identification of flowers in Realtime. This project can be used in Realtime to identify the species of flowers digitally by just from images of flowers

Bibliography

Rajesh Singh, Anita Gehlot, Sushabhan Choudhury, Bhupendra Singh, "Embedded System based on ATmega Microcontroller-Simulation, Interfacing and Projects", Narosa Publishing House, 2017, ISBN: 978-81-8487-5720.

Rajesh Singh, Anita Gehlot, Bhupendra Singh, Sushabhan Choudhury, "Arduino-Based Embedded Systems: Interfacing, Simulation, and LabVIEW GUI"- CRC Press (Taylor & Francis), 2017, ISBN 9781138060784.

Anita Gehlot, Rajesh Singh, Devender Kumar Saini, Monika Yadav, Bhupendra Singh, "IoT Enabled Smart Microgrid.....Rapid Prototyping", GBS Publication, 2018, ISBN: 978-93-87374-41-6.

Anita Gehlot, Rajesh Singh, Mamta Mittal, Bhupendra Singh, Ravindra Sharma, Vikas Garg, "Hands on approach on Applications of Internet of Things in various Domain of Engineering", International Research Publication House, 2018, ISBN: 978-93-87385-25-3.

Rajesh Singh, Anita Gehlot, Bhupendra Singh, Sushabhan Choudhury, "Internet of Things Enabled Automation in Agriculture", New India Publishing Agency, 2018, ISBN: 9789387973053.

Rajesh Singh, Anita Gehlot, Bhupendra Singh, Bikarama Prasad Yadav, "IoT Enabled Fire Safety and Security Devices for Building", Pen2Print, EduPedia Publications, 2018, ISBN:9789386647979.

Rajesh Singh, Anita Gehlot, Bhupendra Singh, Piyush Kuchhal, "Wireless Methods and Devices for Home Automation" International Research Publication House, 2018, ISBN: 978-93-87388-27-7.

Rajesh Singh, Anita Gehlot, Raghuveer Chimata, Bhupendra Singh, P.S.Ranjith, "Internet of Things in Automotive Industries and Road Safety" River Publishers, 2018, ISBN:9788770220101&e-ISBN:9788770220095

Rajesh Singh, Anita Gehlot, Bhupendra Singh, Sushabhan Choudhury, "Arduino meets MATLAB.... Interfacing, Programs and Simulink", Bentham Science, 2018, ISBN: 978-1-68108-728-3 & e-ISBN: 978-1-68108-727-6

Rajesh Singh, Anita Gehlot, Lovi Raj Gupta, Bhunpendra Singh, and Priyanka Tyagi, "Getting Started for Internet of Things with Launch Pad and ESP8266", River Publishers, 2019, ISBN: 9788770220682, e-ISBN: 9788770220675

Rajesh Singh, Anita Gehlot, Bhunpendra Singh, "Arduino and Scilab based Projects", Bentham Science, 2019, ISBN: 9789811410918 & e-ISBN: 9789811410925

Anita Gehlot, Rajesh Singh, Lovi Raj Gupta, Bhupendra Singh, "Cook Book for Mobile Robotic Platform Control with Internet of Things and Ti Launch Pad", BPB Publisher, 2019, ISBN: 978-93-8851-167-4

Rajesh Singh, Anita Gehlot, Lovi Raj Gupta, Bhupendra Singh, Mahindra Swain, "Internet of Things with Raspberry Pi and Arduino", CRC press/Taylor & Francis, 2019, ISBN: 9780367248215

Anita Gehlot, Rajesh Singh, Gaurav Sethi, Bhupendra Singh, "The Robotic Platform Control with Atmega328 and NuttyFi Based on the Internet of Things", Nova Science Publisher, 2020, ISBN:9781536174724

Rajesh Singh, Anita Gehlot, Lovi Raj Gupta, Navjot Rathour, Mahendra Swain, Bhupendra Singh, "IoT based projects", BPB publication, 2020, ISBN: 9789389328523

Chapter 11

Precision Farming

In this Chapter the precision farming/ Agriculture is discussed along with its steps of implementation. It also includes how artificial Intelligence plays role to perform precision farming/ agriculture in various way along with its scope, limitation and challenges.

Introduction

The primary source of livelihood is agriculture in India, where two thirds of India's population, services and the private sector account for the remainder. Agricultural land accounts for about 43 percent of the geographical area of India. In earlier days, India was largely dependent in terms of food imports, but over the years, the country's production of grain and seeds was self-sufficient, and this effort led to the formation of the Green Revolution. In fact, there was strong effort to achieve food self-sufficiency. This trend has continued up to now and continuous improvements have occurred. Precision agriculture or precision farming means the right thing to do, in the right way, on the right place and at the right time. Precision agriculture is expected to suit the agro-climate activities to improve application precision. Farming land has declined a bit in the last 40 years but the number of farmers has increased. According to the 2010–11 Agriculture Census, a total of 138.35 million operating holdings were estimated (single farmers) and 159,59 million hectares were covered by operations.

Precision agriculture is a management technique that collects, processes and analyzes temporary, spatial and individual data and integrates this with other information in order to support management decisions based on an estimated variability to enhance efficiency of use of agricultural production resources, productivity, quality, profitability and sustainability. Precision Agriculture manages every input of crop production (fertilizer, specifically found calestone, herbicide, seeds, insecticide, etc. Reduce waste, boost income and maintain efficiency Place. Precision Agriculture adapts soil and crop management carefully to suit the different field conditions.

Precision Farming Concept

PFS is focused on spatial and temporal variation recognition in crop development. In farm management, variability is taken into account in order to increase productivity and reduce environmental risks. Farms are often large and cover several areas in developed countries. Therefore, spatial variability in large farms has two components: internal variability and interface variability. The precise agriculture program in a field is often called site-specific cultivation management (SSCM).

SSCM refers to an agricultural management system developed that promotes variable practices in a site- or soil-based field

Nevertheless, SSCM is no single technology, according to Batte and VanBuren (1999), but an aggregation of technologies that allows for:

- Data collection on a suitable scale at an appropriate time;
- Data interpretation and analysis to support a range of management choices;
- Applying a management response to an appropriate level at the proper time. Segarra (2002) in a PFS study in developed countries

Emphasizes for farmers the following benefits:

- **Overall increase of yield:** precise crop selection, the application of exact fertilizer types and doses and of herbicides as well as appropriate irrigation meet crop requirements for optimal growth and growth. This leads to increased yields, particularly in areas or fields with historically stable crop management practices.
- **Improved efficiency:** Advanced machinery, tools and information, support farmers in improving labor efficiency, land and time in agriculture. In the US, 1 ha of wheat or corn will be grown in just 2 hours.
- **Reduced costs of production:** The application of exact amounts reduces the costs of agrochemical inputs in crop production at the appropriate time. Furthermore, the overall high return reduces the cost per output unit.
- **Decision-making in agricultural management:** Agricultural machinery, equipment, and devices are used in farmers to obtain reliable information that is processed and evaluated for appropriate decision makers when planting, seeding, fertilizing, applying pesticides and herbicides, irrigating, draining and post-producting.
- **Reduced effect on climate:** The timely application at a appropriate rate of agrochemicals avoids excessive soil and water residues and decreases contamination in the natural area.
- **Accumulation of farmers' awareness to improve management:** All of PFS operations in the field yield useful information on the field and the management and the data is preserved in equipment and computers.

Steps in Precision Farming

The basic steps in precision farming are

Assessing variation

The first critical step in precision farming is the assessment of variability. Because it is obvious that what you don't know cannot be managed. In terms of yield, the factors and processes that regulate or influence crop output vary over time. The precision agriculture challenge is to quantify the variability of these factors and processes, and to decide when and where the spatial and temporal change in crop yield is responsible for different combinations.

Spatial variability measurement methods are readily available and are commonly used in precise agriculture. The most important part of precision farming lies in spatial variability assessment. There are also methods for assessing temporal variability, but spatial and temporal variation are rarely recorded simultaneously. Space and time statistics are important for us. We can see the variability in crop yield in space, but we are evolving during the growing season, which is nothing other than the time variance. Therefore, we need both space-time and precise agricultural statistics. When it come to these factors artificial Intelligence can play a huge role for acquiring all these factors with better accuracy and precision. Artificial Intelligence have various branches like machine learning, deep learning, computer vision and other technologies like data science and bigdata which can play a huge role in collection of data from farming lands to carry out precision farming very easily.

Managing variation

When variability has been properly analyzed, farmers have to align agricultural inputs with established conditions and suggestions on management. They are site-specific and use exact control equipment for applications. We should allow the most use of the technology. Regulation of site-specific variability. We should use the artificial Intelligence system so that we have pronounced the accuracy of the website and simple management. We will remember the sample site co-ordinates and use them for management when taking the soil and plant samples. It contributes to the efficient use of resources and avoids waste, which is what we want and in addition to this with the help of artificial Intelligence when can deploy various bots in the large field for field management and monitoring purpose.

The potential for improved soil fertility management accuracy and increased application control precision make the management of soil fertility an interesting but largely unproven alternative to standardized field management which can be improved with the help of artificial intelligence. The precise soil fertility management definition involves in-field variability and is correctly defined and accurately interpreted to influence crop growth, crop quality and the climate. Inputs may also be precisely applied. The greater the spatial dependence of a sustainable soil, the greater the potential and its potential importance for precision management.

Evaluation

The most important aspect of the study of the profitability of precision farming is that interest is often derived for the use of precision farming through the application of the data and not through use of technology Future changes to environmental quality. The potential benefits to the environment are often cited as reduced agrochemical usage, increased nutrients efficiency, increased input output controlled and increased soil production from degradation. Technologies that allow precision agriculture can be made feasible, agricultural principles and policy rules can make it practical and increase production efficiency or other types of value.

The term technology transfer may mean that precision farming occurs when individuals or companies actually obtain and use the technology that enables them. Although precision agriculture includes using technology to control spatial and temporal variability as well as agronomic concepts, the main word is management. In what is known as technology transfer, much attention has been focusing on how the farmer can interact. Such problems relating to operator managerial capacity, spatial infrastructure delivery and connectivity between the various farms will fundamentally change as precision farming continues to evolve.

Artificial Intelligence in Precision Farming

As we have discussed about the precision farming and its importance in the field of agriculture now let's see how artificial Intelligence can be used at different stages for carrying out precision farming in the agricultural land.

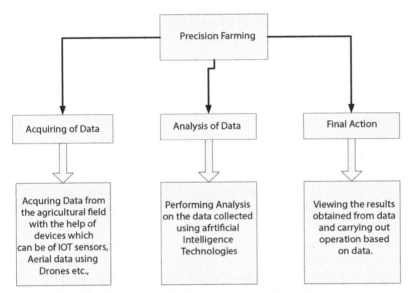

Implementation of Artificial Intelligence in Precision Farming

It can be carried out in 3 steps which are

- **Acquiring of Data**

Acquiring or collecting of data using various technologies like IOT, Drones, Satellite Imagining etc.,

Let's see about these technologies:

IOT in agriculture: The Internet of things is an electronically connected network of physical objects for data collection and aggregation. IoT is part of the sensors and agricultural software creation process. For liquid manure for example, farmers may spectroscopic ally calculate a substantially inconsistent sample of nitrogen, phosphorous and potassium. They can then scan the field to detect the urination of the cows and add fertilizer only to those spots which are essential. This decreases the use of fertilizers by up to 30%. Soil moisture sensors are the best times for water plants remotely.

Drone and Satellite Imaging Technology: Progress in drone and satellite technology benefits precise agriculture, as drones capture images of great quality, while satellites capture the broader picture. Light flight pilots may combine aerial imagery with satellite data to forecast future yields on the basis of current field biomass levels. Aggregate images may create contour maps to track the flow of water, evaluate the rate of seeding and create return maps for areas more or less efficient.

Robots: There have been self-management tractors for some time now as the device works like an autopilot aircraft. The tractor works best, with the farmer in an emergency. Technology is evolving into GPS-programmed driverless machinery for spreading fertilizer or plowing land. Additional innovations are solar energy machines that identify weeds and accurately kill them with an herbicide or laser dose. There are already agricultural robots, also known as Agricultural Bots, but advanced harvesting robots are being built which uses artificial Intelligence in them to detect mature fruit, adjust it to shape and size and carefully pluck it from branches.

- Analysis of Data

After collecting the data from agriculture land at a particular database we can perform various analysis for on the data using various technologies like artificial Intelligence, Machine learning, Deep learning, computer vision etc., from the analysis we can get various valuable factors which can guide us to perform several task precisely to carry out farming in the agricultural field.

- Final Action:

This step involves using various analysis and data   which have collected to make various application based on them which can help to carry out precision farming in the agricultural land very easily. As now day artificial intelligence is involved more into this for making so many applications which can help any formers to perform precision farming very easily by analyzing their crops in agricultural field. The applications which can be made on basis of artificial Intelligence are

- Agriculture Bots

The majority of companies are now preparing and building robots for the main agricultural mission. This involves seed processing and work more efficiently than manpower. This is the latest example of agricultural machine learning.

- Species Breeding

Species selection is a repetitive searching method for specific genes, which decide the effectiveness of water and the use of nutrients, climate tolerance, disease resistance, nutrient content or taste. In the analysis of crop production in different climates and developing new technologies, machine-learning in particular, profound learning algorithms requires decades of field data. Based on this data, a statistical model can be developed that predicts which genes are most likely to contribute to a plant's gain.

- Species Recognition

The conventional approach of the human person for plant classification is comparing the color and shape of the leaves, but by using Machine Leaning's statistical analysis can provide quicker and more detailed results by analyzing the morphology of the leaves and provides more details about the leaf properties.

- Yield Production

Yield forecasting is one of the topics of steep-precision agriculture, as it defines yield mapping and prediction, crop demand matching and crop management. state-of - the-art techniques have gone far beyond simple prediction based through historical data, but have involved machine learning techniques to provide knowledge concerning crops, weather and economic circumstances and extensive multidimensional analysis to optimize the return for farmers and the public.

- Disease Detection or furcating of diseases

Machine Learning can be used to detect Plant infections which are typically caused by plague, insects, diseases and, unless managed on time, they reduce productivity on a large scale. Concerning the area grown in acres, the cultivators are tedious to track the crops daily. By implementing Machine Learning here provides the solution to monitor the cultivated area consistently and provides automatic detection of diseases with remote sensing images.

- Crop Quality

Data output can be adequately measured and defined to lift product price and reduce waste. Contrary to human experts, machines may use apparently irrelevant data and interfaces to discover and identify new qualities that play a role in the overall quality of the crops.

- Weed Detection

Weeds are the major threats to crop production, apart from diseases. The main issue in the war against weeds is that they are hard to distinguish and discriminate against crops. Computer vision and ML algorithms are able to enhance weed identification and discrimination at low cost without environmental issues or side effects. The technologies are going to push robots to kill weeds and reduce the need for herbicides in future.

- Soil Management

Soil is a highly diverse natural resource with complex processes and unclear mechanisms for agricultural specialists. Its own temperature can give insights into the impact on territorial returns of climate change. The algorithms of machine learning study the process of evaporation, soil humidity and temperature to understand ecosystem dynamics and the effects of agriculture.

- Water Management

The hydrological, climatological and agricultural balance of water resources in agriculture has a significance. Using more efficiently irrigated systems and a forecast of a daily dewpoint temperature, the most advanced ML applications so far are connected with an estimation of daily, weekly, or monthly stomata, which helps to classify predicted weathers and to estimation the evaporation.

- Livestock Production

Machine learning offers accurate prediction and calculation of farming parameters, in accordance with crop management, to maximize the economic efficiency of farming systems such as livestock production. In order to allow farmers to change diets and conditions, for example, weight forecasting system will be estimating future weight 150 days before the day they slaughter.

Scope and Limitation in Adoption of Precision Farming in India

Precision Farming principles are applicable to all agricultural sectors including animal breeding, fisheries and forestry. Precision Agriculture (PA, for its Spanish initials) is divided into two groups-Soft PA, Hard PA and soft PA Management assessments, rather than statistical or empirical research, based on observations and intuition. Whereas all the latest technology such as IOT, drone, GPS, GIS, VRT etc. are used as hard PA. 96 million farms in India have less than four acres (ha) of land of a total of 105,3 million farms. While only fragmented land is cultivated, India currently produces almost 200 million tons of food grain. The crop yield per hectare must be economic and without environmental damage in order to compete with world growth. In India, the total consumption of fertilizers is 84.3 Kg / ha and must be reduced in line with guidelines from the fertilizer by systematic soils testing and developing nutrient maps (Lal, 2004). Pest control, disease and weed management, along with nutrient zones, also play an important role in high crop yields. It is possible to track and manage the plague and disease at lower costs using advanced technologies. Some states, such as Punjab, Haryana use fertilizers and pesticides in large doses. The state of Punjab, for example, accounts for 1.5 percent of India's total geographical area, but uses 1.38 million tons of NPK fertilizer (almost 10 percent of all Indians consumption) and 60 percent of weedicides used in India. Typical issues in these areas are total land degradation and inefficient use of agricultural production.

Another field in which Precision Framing will allow Indian farmers to plan irrigation more profitably by adjusting timing, quantity and water placing. Mechanizing agriculture helps farmers reduce labor costs and increase their farm precision, including selection of quality seeds, removal of weeds, pesticides and application of fertilizers, harvesting and sorting of crops according to their quality.

In developing countries in general and in India in particular, there are many constraints on the adoption of precision farming. Some of these limits are common to the ones in other areas; however, Indian conditions are specific:

- The users' society and opinions,
- Small size field,
- No success stories,
- Heterogeneity and sector imperfections of crop systems,
- Ownership of land, facilities and institutional limits,
- Local technical skills lack
- Data accessibility, quality and expense.

Challenges

Since recent decades PA has helped to increase the yield of crops, with decreased cost and human effort, however, due to the following reasons or challenges, the adoption of these new technologies is still very limited by farmers.

- Hardware Cost: Precision Farming/Agriculture primarily relates for the measurement of multiples parameters in real time to equipment including sensors, wireless routers, drones, spectral imaging sensors, etc. These sensors have many drawbacks, including high costs for development, maintenance and operation. Many Precision Farming/Agriculture systems are inexpensive and adaptable for limited arable land, e.g. smart irrigation systems requiring low-cost hardware and sensors. Drone-based crop safety monitoring schemes for large arable soils are not feasible because of high deployment costs.

- Change in the weather: Environmental variability is one of the greatest threats to the accuracy of sensor data. Field sensor nodes are sensitive to environmental variations, such as rain, temperature fluctuations, wind speeds, sunlight, etc. The interference caused by atmospheric disruptions in wireless channels will disrupt contact between wireless node and cloud. Also sensitive to weather changes are the satellite and air-borne drone platforms. Air pollution and other natural aerosols impact the images obtained from these platforms. The development of advanced atmospheric correction, cloud detection and sound interpolation technologies is a major challenge that requires the research community to make a strong effort.

- Excess Data Management: Data is always generated by sensors in Precision Farming/Agriculture. Some data security measures must be put in place to ensure the integrity of data, thereby in turn increasing the system cost. The sensor readings must be precise to take appropriate action when and when necessary. The readings can be corrupted by an attacker and the machine will reduce its performance. Precision Farming/Agriculture systems produce enormous amounts of data, requiring sufficient resources to conduct data analysis. Real-time data obtained from sensors that are deployed in a few minutes and high-altitude / low-altitude spectral imagery produces a bulk of the data that increases the storage and processing requirements. The requirements are new software platforms and scalable management facilities for big data sources. In this context, software solution generation focuses on combining data management with IoT comprehensive cloud computing platforms.

- Literacy rate: Literacy is a major factor influencing the PA adoption ratio. Farmers grow crops based on experience in developing countries with high literacy rates. They do not use state-of-the-art agriculture technologies, which lead to production loss. To understand the technology, farmers need to be educated or trust in technical support from a third party. In underdeveloped countries, the Precision Farming/Agriculture is also not very popular due to constraints on resources and education, where the literacy rate is not high.

- Connectivity: The 5 G networks of the next generation can be 100 times faster than 4 G networks, making communication between servers and devices much faster. Furthermore, 5 G can carry far more information than other networks, making it an ideal technology for data conveyed by remote sensors and drones, key tools tested in PA environments. In current applications, the introduction of modern 5 G communication networks is a must where safe and fast data transmission is in order to allow real-time data processing and decision-making support.

- Interoperability: The interoperability of devices due to various digital standards is one of the biggest challenges facing Precision Farming/Agriculture. This lack of interoperability not only hampers the adoption and slowing down of new IoT technologies but hinders productivity gains by smart applications in agriculture. The Precision Farming/Agriculture scenario currently provides the basis for new approaches and protocols to combine various machine communication requirements in order to explore the potential of effective machine-to - machine communication and data sharing.

Summary

Precision agriculture is a modern approach to increasing crop productivity through the use of the most up-to - date technologies, i.e. IoT, cloud computing, AI and machine Learning (ML), Deep Learning (DL), Computer Vision, Drone Technology etc., Precision Farming/Agriculture aims to provide decision support systems that are focused on a number of crop parameters, i.e. soil nutrients, soil water level, wind speed, sunlight intensity, temperature , humidity, chlorophyll content, and others. However, the development and deployment phase of these systems involves several challenges. Since precision Farming/Agriculture has as its principal objective the optimization of resources such as water, pesticides, fertilizer, etc., to produce excess yields in order to optimize resources, in order to enable artificial intelligence to quantify resources needed for healthy crops at any particular stage of growth.